Numerical Linear Algebra

Numerical Linear Algebra

William Layton
Myron Sussman

University of Pittsburgh, USA

W₀ World Scientific

NEW JERSEY · LONDON · SINGAPORE · BEIJING · SHANGHAI · HONG KONG · TAIPEI · CHENNAI · TOKYO

Published by

World Scientific Publishing Co. Pte. Ltd.

5 Toh Tuck Link, Singapore 596224

USA office: 27 Warren Street, Suite 401-402, Hackensack, NJ 07601

UK office: 57 Shelton Street, Covent Garden, London WC2H 9HE

Library of Congress Cataloging-in-Publication Data
Names: Layton, W. J. (William J.), author. | Sussman, Mike Myron, author.
Title: Numerical linear algebra / William Layton, Mike Myron Sussman, University of Pittsburgh.
Description: New Jersey : World Scientific, 2020. | Includes bibliographical references and index.
Identifiers: LCCN 2020025413 | ISBN 9789811223891 (hardcover) |
  ISBN 9789811224843 (paperback) | ISBN 9789811223907 (ebook) |
  ISBN 9789811223914 (ebook other)
Subjects: LCSH: Algebras, Linear. | Numerical analysis.
Classification: LCC QA184.2 .L395 2020 | DDC 518/.43--dc23
LC record available at https://lccn.loc.gov/2020025413

British Library Cataloguing-in-Publication Data
A catalogue record for this book is available from the British Library.

For any available supplementary material, please visit
https://www.worldscientific.com/worldscibooks/10.1142/11926#t=suppl

Desk Editor: Liu Yumeng

Preface

"It is the mark of an educated mind to rest satisfied with the degree of precision that the nature of the subject permits and not to seek exactness when only an approximation is possible."
— Aristotle (384 BCE)

This book presents numerical linear algebra for students from a diverse audience of senior level undergraduates and beginning graduate students in mathematics, science and engineering. Typical courses it serves include:

A one term, senior level class on Numerical Linear Algebra. Typically, some students in the class will be good programmers but have never taken a theoretical linear algebra course; some may have had many courses in theoretical linear algebra but cannot find the on/off switch on a computer; some have been using methods of numerical linear algebra for a while but have never seen any of its background and want to understand why methods fail sometimes and work sometimes.

Part of a graduate "gateway" course on numerical methods. This course gives an overview in two terms of useful methods in computational mathematics and includes a computer lab teaching programming and visualization connected to the methods.

Part of a one term course on the theory of iterative methods. This class is normally taken by students in mathematics who want to study numerical analysis further or to see deeper aspects of multivariable advanced calculus, linear algebra and matrix theory as they meet applications.

This wide but highly motivated audience presents an interesting challenge. In response, the material is developed as follows: Every topic in numerical linear algebra can be presented algorithmically and theoretically and both views of it are important. The early sections of each chapter present the background material needed for that chapter, an essential step since backgrounds are diverse. Next methods are developed algorithmically with examples. Convergence theory is developed and the parts of the proofs that provide immediate insight into why a method works or how it might fail are given in detail. A few longer and more technically intricate proofs are either referenced or postponed to a later section of the chapter.

Our first and central idea about learning is *"to begin with the end in mind"*. In this book the end is to provide a modern understanding of useful tools. The choice of topics is thus made based on utility rather than beauty or completeness. The theory of algorithms that have proven to be robust and reliable receives less coverage than ones for which knowing something about the method can make a difference between solving a problem and not solving one. Thus, iterative methods are treated in more detail than direct methods for both linear systems and eigenvalue problems. Among iterative methods, the beautiful theory of SOR is abbreviated because conjugate gradient methods are a (currently at least) method of choice for solving sparse SPD linear systems. Algorithms are given in pseudocode based on the widely used MATLAB language. The pseudocode transparently presents algorithmic steps and, at the same time, serves as a framework for computer implementation of the algorithm.

The material in this book is constantly evolving. Welcome!

Contents

Chapter 1

Introduction

There is no such thing as the Scientific Revolution, and this is a book about it.

— Steven Shapin, *The Scientific Revolution.*

This book presents numerical linear algebra. The presentation is intended for the first exposure to the subject for students from mathematics, computer science, engineering. Numerical linear algebra studies several problems:

Linear systems: $Ax = b$: Solve the $N \times N$ linear system.

Eigenvalue problems: $A\phi = \lambda\phi$: Find all the eigenvalues and eigenvectors or a selected subset.

Ill-posed problems and least squares: Find a unique *useful* solution (that is as accurate as possible given the data errors) of a linear system that is undetermined, overdetermined or nearly singular with noisy data.

We focus on the first, treat the second lightly and omit the third. This choice reflects the order the algorithms and theory are built, not the importance of the three. Broadly, there are two types of subproblems: small to medium scale and large scale. *"large "* in large scale problems can be defined as follows: a problem is large if memory management and turnaround time are central challenges. Thus, a problem is not large if one can simply call a canned linear algebra routine and solve the problem reliably within time and resource constraints with no special expertise. Small to medium scale problems can also be very challenging when the systems are very sensitive to data and roundoff errors and data errors are significant. The latter is typical when the coefficients and RHS come from experimental data, which always come with noise. It also occurs when the coefficients depend on physical constants which may be known to only one significant digit.

The origin of numerical linear algebra lies in a 1947 paper of von Neumann and Goldstine [von Neumann and Goldstine (1947)]. Its table of contents, given below, is quite modern in all respects except for the omission of iterative methods:

NUMERICAL INVERTING OF MATRICES OF HIGH ORDER
JOHN VON NEUMANN AND H. H. GOLDSTINE

ANALYTIC TABLE OF CONTENTS

PREFACE
CHAPTER I. The sources of errors in a computation

    1.1. The sources of errors.

        (A) Approximations implied by the mathematical model.
        (B) Errors in observational data.
        (C) Finitistic approximations to transcendental and implicit mathematical formulations.
        (D) Errors of computing instruments in carrying out elementary operations: "Noise." Round off errors. "Analogy" and digital computing. The pseudo-operations.

    1.2. Discussion and interpretation of the errors (A)–(D). Stability.
    1.3. Analysis of stability. The results of Courant, Friedrichs, and Lewy.
    1.4. Analysis of "noise" and round off errors and their relation to high speed computing.
    1.5. The purpose of this paper. Reasons for the selection of its problem.
    1.6. Factors which influence the errors (A)–(D). Selection of the elimination method.
    1.7. Comparison between "analogy" and digital computing methods.

CHAPTER II. Round off errors and ordinary algebraical processes.

    2.1. Digital numbers, pseudo-operations. Conventions regarding their nature, size and use: (a), (b).
    2.2. Ordinary real numbers, true operations. Precision of data. Conventions regarding these: (c), (d).
    2.3. Estimates concerning the round off errors:

1.1 Sources of Arithmetical Error

Errors using inadequate data are much less than those using no data at all.

— Babbage, Charles (1792–1871)

On two occasions I have been asked [by members of Parliament], 'Pray, Mr. Babbage, if you put into the machine wrong figures, will the right answers come out?' I am not able rightly to apprehend the kind of confusion of ideas that could provoke such a question.

— Babbage, Charles (1792–1871)

Numerical linear algebra is strongly influenced by the experience of solving a linear system by Gaussian elimination and getting an answer that is absurd. One early description was in von Neumann and Goldstine [von Neumann and Goldstine (1947)]. They gave 4 sources of errors of types A,B,C,D and a model for computer arithmetic that could be used to track the sources and propagation of roundoff error, the error of type D. In order to understand this type of error, it is necessary to have some understanding

Table 1.1   Common precisions for real numbers.

Common name	Bits	Decimal digits	Max exponent
Single precision	32	$\simeq 8$	38
Double precision	64	$\simeq 16$	308
Quadruple precision	128	$\simeq 34$	4931

of how numbers are represented in computers and the fact that computer arithmetic is only a close approximation to exact arithmetic. Integers, for example, are typically represented in a computer in binary form, with a finite number of binary digits (bits), most commonly 32 or 64 bits, with one bit reserved for the sign of the integer. Exceeding the maximum number of digits can result in anomalies such as the sum of two large positive integers being a negative integer.

Real numbers are typically stored in computers in essentially scientific notation, base 2. As with integers, real numbers are limited in precision by the necessity of storing them with a limited number of bits. Typical precisions are listed in Table 1.1. In Fortran, single precision numbers are called "real", double precision numbers are called "double precision", and quadruple and other precisions are specified without special names. In C and related languages, single precision numbers are called "float", double precision numbers are called "double", and quadruple precision numbers are called "long double". In MATLAB, numbers are double precision by default; other precisions are also available when required.

Machine epsilon. The finite precision of computer numbers means that almost all computer operations with numbers introduce additional numerical errors. For example, there are numbers that are so small that adding them to the number 1.0 will not change its value! The largest of these is often called "**machine epsilon**" and satisfies the property that

$$1 + \epsilon = 1$$

in computer arithmetic.[1] This error and other consequences of the finite length of computer numbers are called "roundoff errors". Generation and propagation of these roundoff errors contains some unpleasant surprises. Everyone who writes computer programs should be aware of the possibilities of roundoff errors as well as other numerical errors.

Common sources of numerical errors. The following five types of error are among the most common sources of numerical errors in computer programs.

---

[1]The precise definition of machine epsilon varies slightly among sources. Some include a factor of 2 so machine epsilon represents the smallest number that changes 1 when added to it instead of the largest which doesn't change the value of 1.

(1) **Input errors.**
One of the most common source of errors in input errors. Typically, you are faced with a program written using double precision numbers but, no matter how hard you try to increase accuracy, only 2 significant digits of accuracy come out. In this case, one likely culprit is an error early in the program where the various constants are defined.

Example 1.1. Somewhere you might find a statement like:

$$\text{pi} = 3.1416$$
$$\text{pi} = 22.0/7.0 \qquad \textbf{WRONG!}$$

To preserve the program's accuracy π must be input to the full sixteen digit accuracy[2] of a double precision number.

$$\text{pi} = 3.1415926535897932$$

A sneaky way around this is:

$$\text{pi} = 4.0 * \text{atan}(1.0)$$

(2) **Mixed mode arithmetic.**
It is generally true that computer arithmetic between two integers will yield an integer, computer arithmetic between single precision numbers will yield a single precision number, *etc.* This convention gives rise to the surprising result that $5/2 = 2$ in computer integer arithmetic! It also introduces the question of how to interpret an expression containing two or more different types of numbers (called "mixed-mode" arithmetic).

Example 1.2. Suppose that the variable X is a single precision variable and the variable Y is a double precision variable. Forming the sum (X + Y) requires first a "promotion" of X temporarily to a double precision value and then adding this value to Y. The promotion does not really add precision because the additional decimal places are arbitrary. For example, the single precision value 3.1415927 might be promoted to the double precision value 3.1415927000000000. Care must be taken when writing programs using mixed mode arithmetic.

Another error can arise when performing integer arithmetic and, especially, when mixing integer and real arithmetic. The following Fortran example program seems to be intended to print the value 0.5, but it

---

[2]In C, numbers are assumed to be double precision, but in Fortran, numbers *must* have their precision specified in order to be sure of their precision.

will print the value 0.0 instead. Analogous programs written in C, C++ and Java would behave in the same way. An analogous MATLAB program will print the value 0.5.

As an example of mixed mode arithmetic, consider this Fortran program.

Example 1.3.

```
integer j,k
real x
j=1
k=2
x=j/k
print *,x
end
```

This program will first perform the quotient 1/2, which is chopped to zero because integer division results in an integer. Then it will set x=0, so it will print the value 0.0 even though the programmer probably expected it to print the value 0.5.

A good way to cause this example program to print the value 0.5 would be to replace the line x=j/k with the line x=real(j)/real(k) to convert the integers to single precision values before performing the division. Analogous programs written in C, C++ and Java can be modified in an analogous way.

(3) **Subtracting nearly equal numbers.**

This is a frequent cause of roundoff error since subtraction causes a loss of significant digits. This source arises in many applications, such as numerical differentiation.

Example 1.4. For example, in a 4-digit mantissa base 10 computer, suppose we do

$$1.234 \times 10^1 - 1.233 \times 10^1 = 1.000 \times 10^{-3}.$$

We go from four significant digits to one. Suppose that the first term is replaced with 1.235×10^1, a difference of approximately 0.1%. This gives

$$1.235 \times 10^1 - 1.233 \times 10^1 = 2.000 \times 10^{-3}.$$

Thus, a 0.1% error in 1.234×10^1 can become a 100% error in the answer!

(4) **Adding a large number to a small one.**
This causes the effect of the small number to be completely lost. This can have profound effects when summing a series of applying a method like the trapezoid rule to evaluate an integral numerically.

Example 1.5. For example, suppose that in our 4-digit computer we perform

$$X = .1234 * 10^3 + .1200 * 10^{-2}.$$

This is done by making the exponents alike and adding the mantissas:

$$.1234 * 10^3$$

\ **CHOP** to 4 digits

$$+ \quad .0000/01200 * 10^3$$

\ **OR ROUND** to 4 digits

$$= 0 .1234 * 10^3.$$

Thus the effect of the small addend is lost on the calculated value of the sum.

(5) **Dividing by a small number.**
This has the effect of magnifying errors: a small percent error can become a large percent error when divided by a small number.

Example 1.6. Suppose we compute, using four significant digits, the following:

$$x = A - B/C,$$

where

$$A = 0.1102 \times 10^9,$$
$$B = 0.1000 \times 10^6,$$
$$C = 0.9000 \times 10^{-3}.$$

We obtain $B/C = .1111 \times 10^9$ and $x = 0.9000 \times 10^6$.
Suppose instead that there is a 0.01% error in calculating C, namely

$$C = 0.9001 \times 10^{-3}.$$

Then we calculated instead

$$B/C = 0.1110 \times 10^9 \quad \text{so} \quad x = 0.1000 \times 10^7.$$

Thus we have an 11% error in the result!

Testing before division. When writing a computer program, in cases where a denominator value can possibly be unrealistically smaller than the numerator, it should be tested before doing the division. For example, by choosing a tiny value appropriate to the quotient at hand, possibly a small multiple of machine epsilon, and testing in the following manner:

```
tiny=10 * machine epsilon
if |denominator| < tiny * | numerator|
  error('Division by zero.')
else
  x=numerator/denominator
end
```

1.2 Measuring Errors: The Trademarked Quantities

Mathematics is not a careful march down a well-cleared highway, but a journey into a strange wilderness, where the explorers often get lost. Rigor should be a signal to the historian that the maps have been made, and the real explorers have gone elsewhere.

— Anglin, W.S. in: "Mathematics and History", Mathematical Intelligencer, v. 4, no. 4.

Since every arithmetic operation induces roundoff error it is useful to come to grips with it on a quantitative basis. Suppose a quantity is calculated by some approximate process. The result, x_{computed}, is seldom the exact or true result, x_{true}. Thus, we measure errors by the following convenient standards. These are "trademarked" terms.

Definition 1.1. Let $\|\cdot\|$ denote a norm. Then

$$\vec{e} = \text{ the error } := \vec{x}_{\text{TRUE}} - \vec{x}_{\text{COMPUTED}},$$

$$e_{\text{ABSOLUTE}} = \|\vec{x}_{\text{TRUE}} - \vec{x}_{\text{COMPUTED}}\|,$$

$$e_{\text{RELATIVE}} = \|\vec{x}_{\text{TRUE}} - \vec{x}_{\text{COMPUTED}}\|/\|\vec{x}_{\text{TRUE}}\|,$$

$$e_{\text{PERCENT}} = \vec{e}_{\text{RELATIVE}} * 100.$$

We generally have:

- **Error:** *essential but unknowable. Indeed, if we know the error and the approximate value, adding then gives the true value. If the true value* *were knowable, then we wouldn't be approximating to start with. If x is a vector with 100,000 components then the error has 100,000 numbers and is also thus beyond understanding in most cases.*

- **Absolute error:** *This replaces many numbers with one number: the magnitude of the error vector. If the absolute error is reduced from 1.0 to .001, then we know for sure that the approximation is improved. This is why we mostly look at error magnitudes and not errors.*

- **Relative error:** *An absolute error of 0.2 might be very bad or very good depending on how big the true solution is. The relative error calibrates the error against the true solution. If the relative error is 10^{-5} then the approximation has 5 significant digits of accuracy.*

- **Percent error:** *This gives another way to think of relative errors for those comfortable with percentages.*

Of course, we seldom know the true solution so it is useful to get a "ballpark" estimate of error sizes. Here are some universally standard ways to estimate roundoff errors:

(1) **(Experimental roundoff errors test)** repeat the calculation in higher precision. The digit where the two results differ represents the place where roundoff error has influenced the lower precision calculation. This also gives an estimate of *how many digits are lost in the lower precision calculation.* From that one estimates how many are lost in higher precision and thus how many to believe are correct.

(2) **(Estimating model errors in the arithmetic model)** Solve the problem at hand twice-once with a given model and second with a more "refined" or accurate arithmetic model. the difference between the two can be taken as a ballpark measure for the error in the less accurate discrete model.

(3) **(Interval Arithmetic for estimating roundoff and other errors)** As a calculation proceeds, we track not only the arithmetic result but also a "confidence interval" is predicted via a worse case type of calculation at every step. Unfortunately, for long calculations, interval arithmetic often gives a worst case confidence interval so wide that it is not very useful.

(4) **(Significant Digit Arithmetic)** Similarly to Interval Arithmetic, the number of significant digits is tracked through each computation.

(5) **(Backward error analysis for studying sensitivity of problem to roundoff error)** For many types of computations, it has been shown

rigorously that "*the solution computed using finite precision is precisely the exact solution in exact arithmetic to a perturbation of the original problem*". Thus the sensitivity of a calculation to roundoff error can be examined by studying the sensitivity of the continuous problem to perturbations in its data.

Exercise 1.1. *What are the 5 main causes of serious roundoff error? Give an example of each.*

Exercise 1.2. *Consider approximating the derivative $f'(a)$ by*

$$f'(a) \approx [f(a+h) - f(a)]/h$$

for h small. How can this introduce serious roundoff error?

Chapter 2

Linear Systems and Finite Precision Arithmetic

"Can you do addition?" the White Queen asked. "What's one and one and one and one and one and one and one and one and one and one?" "I don't know," said Alice. "I lost count."
— Lewis Carroll, Through the Looking Glass.

2.1 Vectors and Matrices

So as they eddied past on the whirling tides,
I raised my voice: "O souls that wearily rove,
Come to us, speak to us — if it be not denied."
Dante Alighieri, L'Inferno, Canto V c. 1300,
(translation of Dorothy L. Sayers).

A **vector** is an ordered collection of n real numbers, an n-tuple:

$$\vec{x} = (x_1, x_2, \ldots, x_n)^t = \begin{bmatrix} x_1 \\ \vdots \\ x_n \end{bmatrix}.$$

Vectors are often denoted by an over arrow, by being written in bold or (most commonly herein) understood from the context in which the vector is used. A **matrix** is a rectangular array of real numbers

$$A_{m \times n} = \begin{bmatrix} a_{11} & a_{12} & \ldots & a_{1n} \\ \vdots & \vdots & \ddots & \vdots \\ a_{m1} & a_{m2} & \ldots & a_{mn} \end{bmatrix}.$$

13

The **transpose of a matrix,** denoted A^t is an $n \times m$ matrix with the rows and columns of A interchanged

$$(A^t)_{n \times m} = \begin{bmatrix} a_{11} & \cdots & a_{m1} \\ a_{12} & \cdots & a_{m1} \\ \vdots & \ddots & \vdots \\ a_{1n} & \cdots & a_{mn} \end{bmatrix}.$$

In other words, if a matrix $A_{m \times n} = (a_{ij})_{\substack{i=1,\ldots,m \\ j=1,\ldots,n}}$ then its transpose $(A^t)_{m \times n} = (a_{ji})_{\substack{j=1,\ldots,n \\ i=1,\ldots,m}}$. For example,

$$\begin{bmatrix} 1 & 2 & 3 \\ 4 & 5 & 6 \end{bmatrix}^t = \begin{bmatrix} 1 & 4 \\ 2 & 6 \\ 3 & 6 \end{bmatrix}.$$

Vector operations of scalar multiplication and vector addition are defined so that vector addition is equivalent to forming the resultant of the two (force) vectors by the parallelogram rule. Thus, if the vectors x, y represent forces, the sum $x + y$ is defined so that $x + y$ is the resultant force of x and y. Conveniently, it means componentwise addition.

Definition 2.1. If $\alpha \in \mathbb{R}$ and $\vec{x}, \vec{y}$ are vectors

$$\alpha \vec{x} = (\alpha x_1, \alpha x_2, \ldots, \alpha x_n)^t,$$
$$\vec{x} + \vec{y} = (x_1 + y_1, x_2 + y_2, \ldots, x_n + y_n)^t.$$

Vector addition and scalar multiplication share many of the usual properties of addition and multiplication of real numbers. One of the most important vector operations is the dot product or the scalar product of two vectors.

Definition 2.2. Given vectors $\vec{x}, \vec{y}$, the dot product or scalar product is the real number

$$\left. \begin{matrix} \vec{x} \cdot \vec{y} \\ \text{or} \\ \langle \vec{x}, \vec{y} \rangle \\ \text{or} \\ (\vec{x}, \vec{y}) \end{matrix} \right\} := x_1 y_1 + x_2 y_2 + \ldots + x_n y_n$$

and the usual (euclidean) length of the vector x is

$$||x||_2 = \sqrt{\vec{x} \cdot \vec{x}} = \sqrt{x_1^2 + x_2^2 + \ldots + x_n^2}.$$

With the definition of matrix multiplication (below) the dot product can also be written $x \cdot y = x^t y$. Recall that the dot product is related to the angle[1] θ between two vectors by the formula

$$\cos \theta = \frac{\langle x, y \rangle}{||x||_2 ||y||_2}.$$

Actually, this formula shows that *as long as any two quantities of the three* (θ, the dot product $\langle \cdot, \cdot \rangle$ and the length $|| \cdot ||_2$) are defined the third is completely determined by the formula. Thus, existence of a dot product is equivalent to being able to define angles between vectors.

If a linear system is to be equivalent to writing *Matrix A times vector x = vector b*, then there is only one consistent way to define the matrix-vector product. Matrix vector products are row × column. This means that the ith component of Ax is equal to (row i of A) *dot product* (the vector x). Matrix matrix multiplication is a direct extension of matrix-vector multiplication.

Definition 2.3. If $A_{m \times n}$ is a matrix and $x_{n \times 1}$ is an n-vector, the product Ax is an m-vector given by

$$(Ax)_i := \sum_{j=1}^{n} a_{ij} x_j.$$

If y is an $m \times 1$ vector we can multiply $y^t A$ to obtain **the transpose of** an n-vector given by

$$\left(y^t A\right)_j = \sum_{i=1}^{n} a_{ij} y_i.$$

Matrix multiplication is possible for matrices of compatible sizes. Thus we can multiply AB if the number of columns of A equals the number of rows of B:

$$A_{m \times n} B_{n \times p} = C_{m \times p}$$

and, in this case,

$$(AB)_{ij} := \sum_{\ell=1}^{n} A_{i\ell} B_{\ell j}, \qquad i = 1, \ldots, m, \quad j = 1, \ldots, p.$$

In words this is:

The i,j entry in AB is the dot product: (The ith row vector in A)· (the jth column vector in B).

---

[1]The same formula is also interpreted as the correlation between x and y, depending on intended application.

For example, a pair of linear systems can be combined into a single system.

$$A_{N \times N} x = b$$

$$\Leftrightarrow$$

$$\begin{bmatrix} a_{11} & a_{12} & \dots & a_{1n} \\ \vdots & \vdots & \ddots & \vdots \\ a_{n1} & a_{n2} & \dots & a_{nn} \end{bmatrix} \begin{bmatrix} x_1 \\ \vdots \\ x_n \end{bmatrix} = \begin{bmatrix} b_1 \\ \vdots \\ b_n \end{bmatrix}$$

and

$$A_{N \times N} y = c$$

$$\Leftrightarrow$$

$$\begin{bmatrix} a_{11} & a_{12} & \dots & a_{1n} \\ \vdots & \vdots & \ddots & \vdots \\ a_{n1} & a_{n2} & \dots & a_{nn} \end{bmatrix} \begin{bmatrix} y_1 \\ \vdots \\ y_n \end{bmatrix} = \begin{bmatrix} c_1 \\ \vdots \\ c_n \end{bmatrix}$$

can be combined into the single, block system

$$\begin{bmatrix} a_{11} & a_{12} & \dots & a_{1n} \\ \vdots & \vdots & \ddots & \vdots \\ a_{n1} & a_{n2} & \dots & a_{nn} \end{bmatrix} \begin{bmatrix} x_1 & y_1 \\ \vdots & \vdots \\ x_n & y_n \end{bmatrix} = \begin{bmatrix} b_1 & c_1 \\ \vdots & \vdots \\ b_n & c_n \end{bmatrix}.$$

Often this is written as

$$AX = B \text{ where } X := [x|y]_{n \times 2}, B = [b|c]_{n \times 2}.$$

In sharp contrast with multiplication of real numbers, multiplication of a pair of $N \times N$ matrices is generally not commutative!

Exercise 2.1.

a. *Pick two 2×2 matrices A, B by filling in the digits of your phone number. Do the resulting matrices commute? Test if the matrices commute with their own transposes.*

b. *[a more advanced exercise] Do a literature search for conditions under which two $N \times N$ matrices commute. If the entries in the matrices are chosen at random, what is the probability they commute? This can be calculated for the 2×2 case directly.*

Exercise 2.2. *Let $x(t)$, $y(t)$ be N vectors that depend smoothly on t. $g(t) := x(t) \cdot y(t)$ is a differentiable function : $R \to R$. By using the* definition of derivative and dot product prove the versions of the product rule of differentiation

$$g'(t) = x'(t) \cdot y(t) + x(t) \cdot y'(t).$$

Exercise 2.3. *Pick two (nonzero) 3-vectors and calculate $x^t y$ and $x y^t$. Notice that the first is a number while the second is a 3×3 matrix. Show that the dimension of the range of that matrix is, aside from special cases where the range is just the zero vector, 1.*

Exercise 2.4. *Find two 2×2 matrices A and B so that $AB = 0$ but neither $A = 0$ nor $B = 0$.*

Exercise 2.5. *Let $x(t), y(t)$ be N vectors that depend smoothly on t. For A an $N \times N$ matrix $g(t) := x(t)^t A y(t)$ is a differentiable function : $R \to R$. Prove the following version of the product rule of differentiation*

$$g'(t) = x'(t)^t A y(t) + x(t)^t A y'(t).$$

2.2 Eigenvalues and Singular Values

"...treat Nature by the sphere, the cylinder and the cone ..."
— Cézanne, Paul (1839–1906)

One of the three fundamental problems of numerical linear algebra is to find information about the eigenvalues of an $N \times N$ matrix A. There are various cases depending on the structure of A (large and sparse vs. small and dense, symmetric vs. non-symmetric) and the information sought (the largest or dominant eigenvalue, the smallest eigenvalue vs. all the eigenvalues).

Definition 2.4 (eigenvalue and eigenvector). Let A be an $N \times N$ matrix. The complex number λ is an **eigenvalue** of A if there is **at least one** nonzero, possibly complex, vector $\vec{\phi} \neq 0$ with

$$A\vec{\phi} = \lambda \vec{\phi}.$$

$\vec{\phi}$ is an eigenvector associated with the eigenvalue λ. The **eigenspace** of λ is the set of all linear combinations of eigenvectors of that λ.

Calculating λ, ϕ by hand (for small matrices typically) is a three step process which is simple in theory but seldom practicable.

Finding λ, $\vec{\phi}$ for an $N \times N$ real matrix A by hand:

- **Step 1:** Calculate exactly the characteristic polynomial of A. $p(\lambda) :=$ $\det(A - \lambda I)$ is a polynomial of degree N with real coefficients.
- **Step 2:** Find the N (counting multiplicities) real or complex roots of $p(\lambda) = 0$. These are the eigenvalues

$$\lambda_1, \lambda_2, \lambda_3, \cdots, \lambda_N$$

- **Step 3:** For each eigenvalue λ_i, use Gaussian elimination to find a non-zero solution of

$$[A - \lambda_i I]\vec{\phi}_i = 0, i = 1, 2, \cdots, N$$

Example 2.1. Find the eigenvalues and eigenvectors of the 2×2 matrix

$$A = \begin{bmatrix} 1 & 1 \\ 4 & 1 \end{bmatrix}.$$

We calculate the degree 2 polynomial

$$p_2(\lambda) = \det(A - \lambda I) = \det \begin{bmatrix} 1 - \lambda & 1 \\ 4 & 1 - \lambda \end{bmatrix} = (1 - \lambda)^2 - 4.$$

Solving $p_2(\lambda) = 0$ gives

$$p_2(\lambda) = 0$$
$$\Leftrightarrow$$
$$(1 - \lambda)^2 - 4 = 0$$
$$\Leftrightarrow$$
$$\lambda_1 = 3, \lambda_2 = -1.$$

The eigenvector $\vec{\phi}_1$ of $\lambda_1 = 3$ is found by solving

$$(A - \lambda I) \begin{bmatrix} x \\ y \end{bmatrix} = \begin{bmatrix} 0 \\ 0 \end{bmatrix}$$
$$\Leftrightarrow$$
$$\begin{bmatrix} -2 & 1 \\ 4 & -2 \end{bmatrix} \begin{bmatrix} x \\ y \end{bmatrix} = \begin{bmatrix} 0 \\ 0 \end{bmatrix}.$$

Solving gives

$$y = t, \qquad -2x + y = 0, \text{ or}$$
$$x = \frac{1}{2}t, \text{ for any } t \in \mathbb{R}.$$

Thus, $(x, y)^t = (\frac{1}{2}t, t)^t$ for any $t \neq 0$ is an eigenvector. For example, $t = 2$ gives

$$\text{eigenvalue: } \lambda_1 = +3,$$

$$\text{eigenvector: } \overrightarrow{\phi}_1 = \begin{bmatrix} 1 \\ 2 \end{bmatrix}$$

$$\text{eigenspace: } \{t \begin{bmatrix} \frac{1}{2} \\ 1 \end{bmatrix} : -\infty < t < \infty\}.$$

Similarly, we solve for $\overrightarrow{\phi}_2$

$$(A - \lambda I) \begin{bmatrix} x \\ y \end{bmatrix} = \begin{bmatrix} 2 & 1 \\ 4 & 2 \end{bmatrix} \begin{bmatrix} x \\ y \end{bmatrix} = \begin{bmatrix} 0 \\ 0 \end{bmatrix}$$

or $(x, y)^t = (-\frac{1}{2}t, t)^t$. Picking $t = 2$ gives

$$\text{eigenvalue: } \lambda_2 = -1,$$

$$\text{eigenvector: } \overrightarrow{\phi}_2 = \begin{bmatrix} -1 \\ 2 \end{bmatrix}$$

$$\text{eigenspace: } \{t \begin{bmatrix} -1 \\ 2 \end{bmatrix} : -\infty < t < \infty\}.$$

It is sometimes true that there are not N independent eigenvectors.

Example 2.2. Find the eigenvalues and eigenvectors of the 2×2 matrix[2]

$$A = \begin{bmatrix} 2 & 1 \\ 0 & 2 \end{bmatrix}.$$

The characteristic polynomial is given by

$$p(\lambda) = (2 - \lambda)^2$$

and there is a single root $\lambda = 2$ of multiplicity 2. To find one eigenvector $\overrightarrow{\phi}_1$, solve the system

$$(A - \lambda I) \begin{bmatrix} x \\ y \end{bmatrix} = \begin{bmatrix} 0 & 1 \\ 0 & 0 \end{bmatrix} \begin{bmatrix} x \\ y \end{bmatrix} = \begin{bmatrix} 0 \\ 0 \end{bmatrix}.$$

All solutions of this system of equations satisfy $x = 0$ with y arbitrary. Hence an eigenvector is given by

$$\overrightarrow{\phi}_1 = \begin{bmatrix} 0 \\ 1 \end{bmatrix}.$$

A second eigenvector, $\overrightarrow{\phi}_2$, would satisfy the same system, so there is no linearly independent second eigenvector!

---

[2]This matrix is easily recognized as a Jordan block.

Example 2.3. Let

$$A = \begin{bmatrix} 0 & 1 \\ -1 & 0 \end{bmatrix}.$$

We calculate as above and find

$$\det[A - \lambda I] = \lambda^2 + 1 = 0$$
$$\lambda_1 = i, \quad \lambda_2 = -i.$$

The eigenvector of $\lambda = +i$ is calculated by Gaussian elimination to be

$$\lambda = +i, \quad \phi = (-i, 1)^T.$$

Exercise 2.6. *Find the eigenvector of $\lambda = -i$.*

Exercise 2.7. *Find the eigenvalues $\lambda(\varepsilon)$ of*

$$A = \begin{bmatrix} +\varepsilon & 1 \\ -1 & -\varepsilon \end{bmatrix}.$$

2.2.1   *Properties of eigenvalues*

Eigenvalues and eigenvectors are mathematically interesting and important because they give geometric facts about the matrix A. Two of these facts are given in the following theorem.

Theorem 2.1. *(i) Let A be an $N \times N$ matrix. If x is any vector in the eigenspace of the eigenvalue λ then Ax is just multiplication of x by λ: $Ax = \lambda x$.*

(ii) A is invertible if and only of no eigenvalue of A is zero.

It is much harder to connect properties of eigenvalues to values of specific entries in A. In particular, the eigenvalues of A are complicated nonlinear functions of the entries in A. Thus, the eigenvalues of $A + B$ can have no general correlation with those of A and B. In particular, eigenvalues are not additive: generally $\lambda(A + B) \neq \lambda(A) + \lambda(B)$.

Another geometric fact is given in the following exercise.

Exercise 2.8. *Given two commuting matrices A and B, so that $AB = BA$, show that if x is an eigenvector of A then it is also an eigenvector of B, but with a possibly different eigenvalue.*

Proposition 2.1 (Eigenvalues of triangular matrices). *If A is diagonal, upper triangular or lower triangular, then the eigenvalues are on the diagonal of A.*

Proof. Let A be upper triangular. Then, using $*$ to denote a generic non-zero entry,

$$\det [A - \lambda I] = \det \begin{bmatrix} a_{11} - \lambda & * & * & * \\ 0 & a_{22} - \lambda & * & * \\ 0 & 0 & \ddots & * \\ 0 & 0 & 0 & a_{nn} - \lambda \end{bmatrix} =$$

expand down column 1 and repeat

$$= (a_{11} - \lambda)(a_{22} - \lambda) \cdot \ldots \cdot (a_{nn} - \lambda) = p_n(\lambda).$$

The roots of p_n are obviously a_{ii}. $\qquad\square$

When the matrix A is symmetric, its eigenvalues and eigenvectors have special, and very useful, properties.

Proposition 2.2 (Eigenvalues of symmetric matrices). *If A is symmetric (and real) $(A = A^t)$, then:*

(i) all the eigenvalues and eigenvectors are real.

(ii) there exists N orthonormal[3] eigenvectors $\vec{\phi}_1, \ldots, \vec{\phi}_N$ of A:

$$\langle \vec{\phi}_i, \vec{\phi}_j \rangle = \begin{cases} 1, \text{ if } i = j, \\ 0, \text{ if } i \neq j. \end{cases}$$

(iii) if C is the $N \times N$ matrix with eigenvector $\vec{\phi}_j$ in the jth column then

$$C^{-1} = C^t \qquad and \qquad C^{-1}AC = \begin{bmatrix} \lambda_1 & 0 & \ldots & 0 \\ 0 & \lambda_2 & \ldots & 0 \\ \vdots & \vdots & \ddots & \vdots \\ 0 & 0 & \ldots & \lambda_N \end{bmatrix}.$$

In the case that A is not symmetric, the eigenvalues and eigenvectors might not be real. In addition, there might be fewer than N eigenvectors.

Example 2.4. The matrix A below has eigenvalues given by $\lambda_1 = +i$ and $\lambda_2 = -i$:

$$A = \begin{bmatrix} 0 & 1 \\ -1 & 0 \end{bmatrix}.$$

---

[3] "Orthonormal" means that the vectors are orthogonal (mutually perpendicular so their dot products give zero) and normal (their lengths are normalized to be one).

For some calculations, the so called **singular values** of a matrix are of greater importance than its eigenvalues.

Definition 2.5 (Singular values). The singular values of a real $N \times N$ matrix A are $\sqrt{\lambda(A^t A)}$.

The square root causes no problem in to the definition of singular values. The matrix $A^t A$ is symmetric so its eigenvalues are real. Further, they are also nonnegative since $A^t A\phi = \lambda\phi$, both λ, ϕ are real and thus

$$\langle A^t A\phi, \phi \rangle = \lambda \langle \phi, \phi \rangle \quad \text{so}$$

$$\lambda = \frac{\langle A^t A\phi, \phi \rangle}{\langle \phi, \phi \rangle} = \frac{\langle A\phi, A\phi \rangle}{\langle \phi, \phi \rangle} = \frac{|A\phi|_2^2}{|\phi|_2^2} \geq 0.$$

Exercise 2.9. *Prove that*[4]

$$\det \begin{bmatrix} a_{11} & * & * & * \\ 0 & a_{22} & * & * \\ 0 & 0 & \ddots & * \\ 0 & 0 & 0 & a_{nn} \end{bmatrix} = a_{11} \cdot a_{22} \cdot \ldots \cdot a_{nn}.$$

Exercise 2.10. *Pick two (nonzero) 3-vectors and calculate the 3×3 matrix xy^t. Find its eigenvalues. You should get 0,0, and something nonzero.*

Exercise 2.11. *Let*

$$A = \begin{bmatrix} 1 & t \\ -t & 1 \end{bmatrix}.$$

Find its eigenvalues and eigenvectors explicitly as a function of t. Determine if they are differentiable functions of t.

2.3   Error and Residual

"The errors of definitions multiply themselves according as the reckoning proceeds; and lead men into absurdities, which at last they see but cannot avoid, without reckoning anew from the beginning."

— Hobbes, Thomas, In J. R. Newman (ed.), The World of Mathematics, New York: Simon and Schuster, 1956.

---

[4]Here "*" denotes a generic non-zero real number. This is a common way to represent the non-zero entries in a matrix in cases where either their exact value does not affect the result or where the non-zero pattern is the key issue.

Numerical linear algebra is concerned with solving the eigenvalue problem $A\vec{\phi} = \lambda\vec{\phi}$ (considered in Chapter 9) and solving the linear system $Ax = b$ (which we begin considering now). Computer solutions for these problems are *always wrong* because we cannot solve either exactly to infinite precision. For the linear system we thus produce an *approximation*, $\hat{x}$, to the *exact* solution, $x = A^{-1}b$. We are concerned, then, with "how wrong" $\hat{x}$ is. Two useful measures are:

Definition 2.6. Let $Ax = b$ and let $\hat{x}$ be any vector. The **error** (vector) is $e := x - \hat{x}$ and the **residual** (vector) is $\hat{r} := b - A\hat{x}$.

Obviously, *the error is zero if and only if the residual is also zero*. Errors and residuals have a geometric interpretation:

The size of the error is a measure of the distance between the exact solution, x and its approximation $\hat{x}$.
The size of the residual is a measure of how close $\hat{x}$ is to satisfying the linear equations.

Example 2.5 (Error and residual for 2×2 systems). Consider the 2×2 linear system

$$x - y = 0$$
$$-0.8x + y = 1/2.$$

This system represents two lines in the plane, plotted in Figure 2.1, and the solution of the system is the intersection of the two lines.

Consider the point $P = (0.5, 0.7)$ which is on the plot as well. The size of the error is the distance from P to the intersection of the two lines. The error is thus relatively large in Figure 2.1. However, the size of the residual is the distance of P to the two lines. For this example, the point P is close to both lines so this is a case where the residual is expected to be smaller than the error.

For general $N \times N$ systems, the error is essential but, in a very real sense unknowable. Indeed, if we knew the exact error then we could recover the exact solution by $x = \hat{x} + e$. If we could find the exact solution, then we wouldn't be approximating it in the first place! The residual is easily computable so it is observable. It also gives some indication about the error as whenever $\hat{r} = 0$, then necessarily $e = 0$. Thus much of numerical linear algebra is about using the observable residual to infer the size of the unknowable error. The connection between residual and error is given

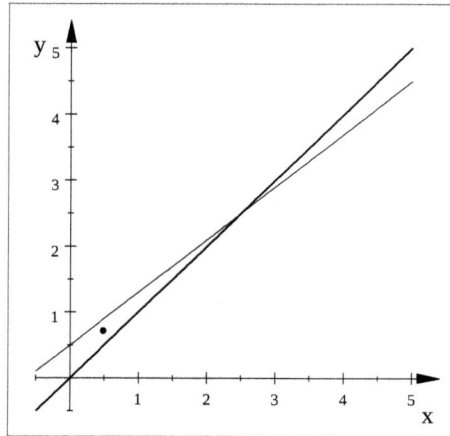

Fig. 2.1   Lines L1, L2 and the point P.

in the following theorem, the Fundamental Equation of Numerical Linear Algebra (FENLA).

Theorem 2.2 (FENLA). *Given a square $N \times N$ linear system $A_{N \times N} x = b$ and $\widehat{x}$. Let $e := x - \widehat{x}$ and $\widehat{r} := \overrightarrow{b} - A\widehat{x}$ be the error and residual respectively. Then*

$$Ae = \widehat{r}.$$

Proof. This is an identity so it is proven by expanding and rearranging:

$$Ae = A(x - \widehat{x}) = Ax - A\widehat{x} = b - A\widehat{x} = \widehat{r}.$$

$\square$

In pursuit of error estimates from residuals, the most common vector and matrix operations include residual calculations, triad calculations, quadratic form calculations, and norm calculations.

Residual Calculation. Given a square $N \times N$ linear system $Ax = b$ and a candidate for its solution N-vectors $\widehat{x}$, compute the residual:

$$\widehat{r} := \overrightarrow{b} - A\widehat{x}.$$

Triad Calculation. Given n-vectors $\overrightarrow{x}, \overrightarrow{y}$ and $\overrightarrow{z}$ compute the vector

$$\overrightarrow{x} + (\overrightarrow{x} \cdot \overrightarrow{y})\overrightarrow{z}.$$

Quadratic Form Calculation. Given a square $N \times N$ matrix $A_{N \times N}$ and n-vectors $\vec{x}$ and $\vec{y}$ compute the *number*

$$\vec{y} \cdot (A\vec{x}) = \vec{y}^{tr} A \vec{x} = \sum_{i,j=1}^{n} y_i A_{ij} x_j.$$

The quadratic form reduces a lot of information ($n^2 + 2n$ real numbers) to one real number.

Norm Calculation. For an n-vector $\vec{x}$ compute norms (weighted averages) such as the RMS (root mean square) norm

$$||\vec{x}||_{\text{RMS}} = \sqrt{\frac{1}{n} \sum_{i=1}^{n} |x_i|^2}.$$

Often, norms of residuals are computed:

$$Step\ 1 : \ \widehat{r} := \vec{b} - A\widehat{x}$$

$$Step\ 2 : \ ||\widehat{r}||_{\text{RMS}} = \sqrt{\frac{1}{n} \sum_{i=1}^{n} |r_i|^2}.$$

This last calculation is an example of a vector norm. In approximating the solution (a vector) to a linear system, the error must be measured. The error is typically measured by an appropriate norm (a generalization of the idea of the length of a vector). Some typical vector norms are given in the following definition.

Definition 2.7 (Three vector norms). Three commonly used vector norms are given as follows.

$$||\vec{x}||_1 := \sum_{i=1}^{n} |x_i|,$$

$$||\vec{x}||_2 := \sqrt{\sum_{i=1}^{n} |x_i|^2},$$

$$||\vec{x}||_\infty := \max_{1 \le i \le n} |x_i|.$$

When solving problems with large numbers of unknowns (large n) it is usually a good idea to scale the answers to be $O(1)$. This can be done by computing relative errors. It is sometimes[5] done by scaling the norms so

---

[5]Computer languages such as MATLAB have built-in functions to compute these norms. These built-in functions do not compute the scaled form.

the vector of all 1's has norm equal to 1 as follows

$$||\vec{x}||_{average} := \frac{1}{n}\sum_{i=1}^{n}|x_i|, \text{ and } ||\vec{x}||_{RMS} := \sqrt{\frac{1}{n}\sum_{i=1}^{n}|x_i|^2}.$$

Exercise 2.12. *Consider the 2×2 linear system with solution $(1,1)$*

$$1.01x + 0.99y = 2$$
$$0.99x + 1.01y = 2.$$

Let the approximate solution be $(2,0)$. Compute the following quantities.

(1) The error vector,
(2) The 2 norm of the error vector,
(3) The relative error (norm of the error vector divided by norm of the exact solution),
(4) The residual vector, and,
(5) The 2 norm of the residual vector.

Exercise 2.13. *Suppose you are given a matrix, A, a right hand side vector, b, and an approximate solution vector, x. Write a computer program to compute each of the following quantities.*

(1) The error vector,
(2) The 2 norm of the error vector,
(3) The relative error (norm of the error vector divided by norm of the exact solution),
(4) The residual vector, and,
(5) The 2 norm of the residual vector.

Test your program with numbers from the previous exercise. **Hint:** *If you are using* MATLAB, *the* **norm** *function can be used to compute the unscaled quantity $|| \cdot ||_2$.*

Exercise 2.14. *Given a point (x_0, y_0) and two lines in the plane. Calculate the distance to the lines and relate it to the residual vector. Show that*

$$||r||_2^2 = (1 + m_1^2)d_1^2 + (1 + m_2^2)d_2^2$$

where m, d are the slopes and distance to the line indicated.

2.4   When is a Linear System Solvable?

Mathematics is written for mathematicians.
— Copernicus, Nicholaus (1473–1543), "De Revolutionibus orbium coelestium"

"Of my 57 years I've applied at least 30 to forgetting most of what I've learned or read, and since I've succeeded in this I have acquired a certain ease and cheer which I should never again like to be without. ... I have stored little in my memory but I can apply that little and it is of good use in many and varied emergencies..."
— Emanuel Lasker

Much of the theory of linear algebra is dedicated to giving conditions on the matrix A that can be used to test if an $N \times N$ linear system

$$Ax = b$$

has a unique solution for every right hand side b. The correct condition is absolutely clear for 2×2 linear systems. Consider therefore a 2×2 linear system

$$\begin{bmatrix} a_{11} & a_{12} \\ a_{21} & a_{22} \end{bmatrix} \begin{bmatrix} x \\ y \end{bmatrix} = \begin{bmatrix} b_1 \\ b_2 \end{bmatrix}.$$

The reason to call the variables x and y (and not x_1, x_2) is that the 2×2 case is equivalent to looking in the $x - y$ plane for the intersection of the two lines (and the "solution" is the $x - y$ coordinates of the intersection point of the 2 lines)

$$\text{Line } L1: \ a_{11}x + a_{12}y = b_1$$
$$\text{Line } L2: \ a_{21}x + a_{22}y = b_2.$$

Plotting two lines in the $x - y$ plane the three possible cases are shown in Figure 2.2.

(a) If L1 and L2 are not parallel, then a unique solution exists for all RHS.
(b) If L1 is on top of L2, than an infinite number of solutions exist for that particular RHS and no solution for any other RHS.
(c) If L1 is parallel to L2 and they are not the same line, then no solution exists. Otherwise, there are an infinite number of solutions.

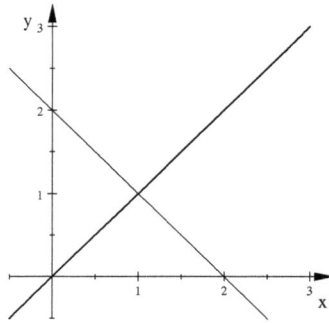

(a) Lines L1, L2 are not parallel: Unique intersection

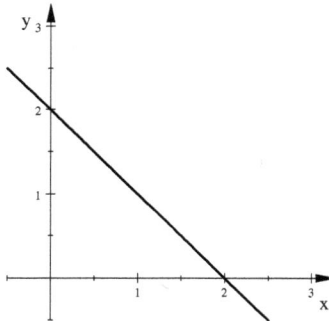

(b) L1: $x + y = 2$, L2: $2x + 2x = 4$ are on top of one another: Infinite number of solutions

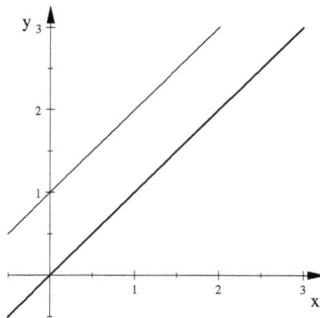

(c) Lines L1, L2 are parallel: No intersection

Fig. 2.2   Three possibilities for two lines.

Unique solvability thus depends on the angle between the two lines: If it is not 0 or 180 degrees a unique solution exists for every possible right hand side.

For the general $N \times N$ linear system, the following is known.

Theorem 2.3 (Unique solvability of $Ax = b$). *The $N \times N$ linear system $Ax = b$ has a unique solution x for every right hand side b if and only if any of the following equivalent conditions holds.*

*1. [**The null space of A is trivial**] The only solution of $Ax = \vec{0}$ is the zero vector, $x = \vec{0}$.*

*2. [**Uniqueness implies existence**] $Ax = b$ has at most one solution for every RHS b.*

*3. [**Existence implies uniqueness**] $Ax = b$ has at least one solution for every RHS b.*

*4. [**A restatement of trivial null space**] The kernel or null space of A is $\{\vec{0}\}$:*

$$N(A) := \{x : Ax = 0\} = \{\vec{0}\}.$$

*5. [**A restatement of existence implies uniqueness**] The range of A is $\mathbb{R}^N$:*

$$Range(A) := \{y : y = Ax \text{ for some } x\} = \mathbb{R}^N.$$

*6. [**Nonzero determinant condition**] The determinant of A satisfies*

$$\det(A) \neq 0.$$

*7. [**Nonzero eigenvalue condition**] No eigenvalue of A is equal to zero:*

$$\lambda(A) \neq 0 \text{ for all eigenvalues } \lambda \text{ of } A.$$

There are many more.

"The well does not walk to the thirsty man."
Transuranian proverb (J. Burkardt)

Exercise 2.15. *Consult reference sources in theoretical linear algebra (books or online) and find 10 more unique solvability conditions.*

2.5 When is an N×N Matrix Numerically Singular?

> To your care and recommendation am I indebted for having replaced a half-blind mathematician with a mathematician with both eyes, which will especially please the anatomical members of my Academy.
>
> — Frederick the Great (1712–1786), [To D'Alembert about Lagrange. Euler had vacated the post.] In D. M. Burton, Elementary Number Theory, Boston: Allyn and Bacon, Inc., 1976.

Many of the theoretical conditions for unique solvability are conditions for which no numerical value can be assigned to see how close a system might be to being singular. The search for a way to quantify how close to singular a system might be has been an important part of numerical linear algebra.

Example 2.6 (Determinant does not measure singularity).
Consider the two lines

$$-x + y = 1 \quad \text{and}$$
$$-1.0000001x + y = 2.$$

Their slopes are $m = 1$ and $m = 1.0000001$. Thus the angle between them is very small and the matrix below must be almost singular

$$\begin{bmatrix} -1 & 1 \\ -1.0000001 & 1 \end{bmatrix}.$$

Many early researchers had conjectured that the determinant was a good measure of this (for example, the above determinant is 0.0000001 which is indeed small). However, multiplying the second equation through by 10^7 does not change the 2 lines, now written as

$$-x + y = 1 \quad \text{and}$$
$$-10000001x + 10000000y = 20000000,$$

or (obviously) the angle between them. The new coefficient matrix is now

$$\begin{bmatrix} -1 & 1 \\ -10000001 & 10000000 \end{bmatrix}.$$

The linear system is still as approximately singular as before but the new determinant is now exactly 1. Thus:

How close det(A) is to zero is not a measure of how close a matrix is to being singular.

Goldstine, von Neumann and Wilkinson found the correct path by looking at 2×2 linear systems (we have been following their example). Consider therefore a 2×2 linear system

$$\begin{bmatrix} a_{11} & a_{12} \\ a_{21} & a_{22} \end{bmatrix} \begin{bmatrix} x \\ y \end{bmatrix} = \begin{bmatrix} b_1 \\ b_2 \end{bmatrix}$$

$$\Leftrightarrow$$

Line $L1$: $\quad a_{11}x + a_{12}y = b_1$

Line $L2$: $\quad a_{21}x + a_{22}y = b_2$.

Plotting two lines in the $x - y$ plane, geometrically it is clear that the right definition for 2×2 systems of *almost singular or numerically singular* is as follows.

Definition 2.8. For the 2×2 linear system above, the matrix A is almost or numerically singular if the angle between the lines L1 and L2 is almost zero or zero to numerical precision.

Exercise 2.16.

(1) For $\varepsilon > 0$ a small number, consider the 2×2 system:

$$x + y = 1,$$

$$(1 - 2\varepsilon)x + y = 2,$$

$$\text{and let } A = \begin{bmatrix} 1 & 1 \\ 1 - 2\varepsilon & 1 \end{bmatrix}.$$

*(2) **Find the eigenvalues** of the coefficient matrix A.*
*(3) **Sketch** the two lines in the $x - y$ plane the system represents. On the basis of your sketch, **explain** if A is ill conditioned and why.*

In the following chapter, numerical methods for solving linear systems are discussed and, along the way, the notion of numerical singularity will be refined and methods to estimate numerical singularity of large systems will be given.

Chapter 3

Gaussian Elimination

> One of the main virtues of an electronic computer from the point
> of view of the numerical analyst is its ability to "do arithmetic fast".
> — James Wilkinson, 1971.

Gaussian elimination is the basic algorithm of linear algebra and the
workhorse of computational mathematics. It is an algorithm for solving
exactly (in exact arithmetic) the $N \times N$ system:

$$A_{N \times N} x_{x_{N \times 1}} = b_{N \times 1}, \qquad \text{where} \quad \det(A) \neq 0. \tag{3.1}$$

It is typically used on all matrices with mostly non-zero entries (so called
dense matrices) and on moderate sized, for example $N \leq 10,000$,[1] matrices
which have only a few non zero entries per row that occur in some regular
pattern (these are called banded and sparse matrices). Larger matrices,
especially ones without some regular pattern of non zero entries, are solved
using iterative methods.

3.1 Elimination + Backsubstitution

> Luck favors the prepared mind.
> — Louis Pasteur

The $N \times N$ system of equations $Ax = b$ is equivalent to

$$
\begin{aligned}
a_{11}x_1 + a_{12}x_2 + \;\ldots\; + a_{1N}x_N &= b_1, \\
a_{21}x_1 + a_{22}x_2 + \;\ldots\; + a_{2N}x_N &= b_2, \\
&\;\;\vdots \\
a_{N1}x_1 + a_{N2}x_2 + \ldots + a_{NN}x_N &= b_N.
\end{aligned}
\tag{3.2}
$$

---

[1]The number $N = 10,000$ dividing small from large is machine dependent and will
likely be incorrect (too small) by a year after these notes appear.

Gaussian elimination solves it in two phases: **elimination** followed by **backsubstitution**.

The Elimination Step: The elimination step reduces the matrix A to upper form by operations which do not alter the solution of (3.1), (3.2). These "*Elementary Row Operations*"[2] are:

(1) Multiply a row of (3.2) by a non-zero scalar.
(2) Add a multiple of one row of (3.2) to another.
(3) Interchange two rows of (3.2).

To show how these are used, consider (3.1):

$$\begin{bmatrix} a_{11} & a_{12} & \ldots & a_{1N} \\ a_{21} & a_{22} & \ldots & a_{2N} \\ \vdots & \vdots & \ddots & \vdots \\ a_{N1} & a_{N2} & \ldots & a_{NN} \end{bmatrix} \begin{bmatrix} x_1 \\ x_2 \\ \vdots \\ x_n \end{bmatrix} = \begin{bmatrix} b_1 \\ b_2 \\ \vdots \\ b_n \end{bmatrix}.$$

Gaussian elimination proceeds as follows.

Substep 1: Examine the entry a_{11}. If it is zero or too small, find another matrix entry and interchange rows or columns to make this the entry a_{11}. This process is called "**pivoting**" and a_{11} is termed the "**pivot entry**". Details of pivoting will be discussed in a later section, so for now, just assume a_{11} is already suitably large.

With the pivot entry non-zero, add a multiple of row 1 to row 2 to make a_{21} zero:

$$\text{Compute:} \quad m_{21} := \frac{a_{21}}{a_{11}},$$
$$\text{Then compute:} \quad Row\ 2 \Leftarrow Row\ 2 - m_{21} \cdot Row\ 1.$$

This zeroes out the $2, 1$ entry and gives

$$\begin{bmatrix} a_{11} & a_{12} & \ldots & a_{1N} \\ 0 & a_{22} - m_{21}a_{12} & \ldots & a_{2N} - m_{21}a_{1N} \\ a_{31} & a_{32} & \ldots & a_{3N} \\ \vdots & \vdots & \ddots & \vdots \\ a_{N1} & a_{N2} & \ldots & a_{NN} \end{bmatrix} \begin{bmatrix} x_1 \\ x_2 \\ x_3 \\ \vdots \\ x_N \end{bmatrix} = \begin{bmatrix} b_1 \\ b_2 - m_{21}b_1 \\ b_3 \\ \vdots \\ b_N \end{bmatrix}.$$

Note that the $2, 1$ entry (and all the entries in the second row and second component of the RHS) are now replaced by new values. Often

---

[2]It is known that operation 1 multiplies $det(A)$ by the scalar, operation 2 does not change the value of $det(A)$ and operation 3 multiplies $det(A)$ by -1.

the replacement is written by an arrow, such as $a_{22} \Leftarrow a_{22} - m_{21}a_{12}$.
Often its denoted by an equals sign. This is not a mathematical equals
sign but really denotes an assignment meaning : "Replace LHS by
RHS" as in $a_{22} = a_{22} - m_{21}a_{12}$. Since this replacement of values is
what is really done on the computer we have the system

$$\begin{bmatrix} a_{11} & a_{12} & \dots & a_{1N} \\ 0 & a_{22} & \dots & a_{2N} \\ a_{31} & a_{32} & \dots & a_{3N} \\ \vdots & \vdots & \ddots & \vdots \\ a_{N1} & a_{N2} & \dots & a_{NN} \end{bmatrix} \begin{bmatrix} x_1 \\ x_2 \\ x_3 \\ \vdots \\ x_N \end{bmatrix} = \begin{bmatrix} b_1 \\ b_2 \\ b_3 \\ \vdots \\ b_N \end{bmatrix}$$

where the second row now contains different numbers than before
step 1.

Substep 1 continued: Continue down the first column, zeroing out the
values below the diagonal (the pivot) in column 1:

$$\text{Compute:} \quad m_{31} := \frac{a_{31}}{a_{11}},$$
$$\text{Then compute:} \quad Row\ 3 \Leftarrow Row\ 3 - m_{31} \cdot Row\ 1,$$

$$\dots \ \dots$$

$$\text{Compute:} \quad m_{N1} := \frac{a_{N1}}{a_{11}},$$
$$\text{Then compute:} \quad Row\ n \Leftarrow Row\ N - m_{N1} \cdot Row\ 1.$$

The linear system now has the structure:

$$\begin{bmatrix} a_{11} & a_{12} & \dots & a_{1N} \\ 0 & a_{22} & \dots & a_{2N} \\ \vdots & \vdots & \ddots & \vdots \\ 0 & a_{N2} & \dots & a_{NN} \end{bmatrix} \begin{bmatrix} x_1 \\ x_2 \\ \vdots \\ x_N \end{bmatrix} = \begin{bmatrix} b_1 \\ b_2 \\ \vdots \\ b_N \end{bmatrix}.$$

Step 2: Examine the entry a_{22}. If it is zero or too small, find another
matrix entry below (and sometimes to the right of) a_{22} and interchange
rows (or columns) to make this entry a_{22}. Details of this pivoting
process will be discussed later.

With the pivot entry non zero, add a multiple of row 2 to row 3 to
make a_{32} zero:

$$\text{Compute } 3, 2 \text{ multiplier:} \quad m_{32} := \frac{a_{32}}{a_{22}},$$
$$\text{Then compute:} \quad Row\ 3 \Leftarrow Row\ 3 - m_{32} \cdot Row\ 2.$$

Step 2 continued: Continue down column 2, zeroing out the values below the diagonal (the pivot):

$$\text{Compute:} \quad m_{42} := \frac{a_{42}}{a_{22}},$$

$$\text{Then compute:} \quad Row\ 4 \Leftarrow Row\ 4 - m_{42} \cdot Row\ 2,$$

$$\cdots \quad \cdots$$

$$\text{Compute:} \quad m_{N2} := \frac{a_{N2}}{a_{22}},$$

$$\text{Then compute:} \quad Row\ N \Leftarrow Row\ N - m_{N2} \cdot Row\ 2.$$

The linear system now has the structure:

$$\begin{bmatrix} a_{11} & a_{12} & \ldots & a_{1N} \\ 0 & a_{22} & \ldots & a_{2N} \\ 0 & 0 & \ddots & \vdots \\ 0 & 0 & \ldots & a_{NN} \end{bmatrix} \begin{bmatrix} x_1 \\ x_2 \\ \vdots \\ x_N \end{bmatrix} = \begin{bmatrix} b_1 \\ b_2 \\ \vdots \\ b_N \end{bmatrix}.$$

Substeps 3 through N: Proceed as above for column 2, for each of columns 3 through N. The diagonal entries $a_{33} \ldots a_{NN}$ become pivots (and must not be too small). When Gaussian elimination terminates, the linear system has the structure (here depicted only for the case $N = 4$, or 4×4 matrix):

$$\begin{bmatrix} a_{11} & a_{12} & a_{13} & a_{14} \\ 0 & a_{22} & a_{23} & a_{24} \\ 0 & 0 & a_{33} & a_{34} \\ 0 & 0 & 0 & a_{44} \end{bmatrix} \begin{bmatrix} x_1 \\ x_2 \\ x_3 \\ x_4 \end{bmatrix} = \begin{bmatrix} b_1 \\ b_2 \\ b_3 \\ b_4 \end{bmatrix}.$$

The Backsubstitution Step: We now have reduced the linear system to an equivalent upper triangular system with the same solution. That solution is now quickly found by back substitution as follows.

Substep 1: $a_{NN} x_N = b_N$ so $x_N = b_N / a_{NN}$

Substep 2: $x_{N-1} = (b_{N-1} - a_{N-1,N} x_N) / a_{N-1,N-1}$

Substep 3: $x_{N-2} = (b_{N-2} - a_{N-2,N-1} x_{N-1} - a_{N-2,N} x_N) / a_{N-2,N-2}$

Substeps 4–(N − 1): Continue as above.

Substep N: $x_1 = (b_1 - a_{12} x_2 - a_{13} x_3 - \cdots - a_{1N} x_N) / a_{11}$.

3.2   Algorithms and Pseudocode

Careful analysis of algorithms requires some way to make them more precise. While the description of the Gaussian elimination algorithm provided

in the previous section is clear and complete, it does not provide a straightforward roadmap to writing a computer program. Neither does it make
certain aspects of the algorithm obvious: for example it is hard to see why
the algorithm requires $O(N^3)$ time for an $N \times N$ matrix.

In contrast, a computer program would provide an explicit implementation of the algorithm, but it would also include details that add nothing to
understanding the algorithm itself. For example, the algorithm would not
change if the matrix were written using single precision or double precision
numbers, but the computer program would. Further, printed computer
code is notoriously difficult for readers to understand. What is needed
is some intermediate approach that marries the structural precision of a
computer program with human language descriptions and mathematical
notation.

This intermediate approach is termed "**pseudocode**". A recent
Wikipedia article[3] describes pseudocode in the following way.

> "Pseudocode is a compact and informal high-level description of
> a computer programming algorithm that uses the structural con-
> ventions of a programming language, but is intended for human
> reading rather than machine reading. Pseudocode typically omits
> details that are not essential for human understanding of the al-
> gorithm, such as variable declarations .... The programming lan-
> guage is augmented with natural language descriptions of the de-
> tails, where convenient, or with compact mathematical notation.
> The purpose of using pseudocode is that it is easier for humans
> to understand than conventional programming language code, and
> that it is a compact and environment-independent description of
> the key principles of an algorithm. It is commonly used in text-
> books and scientific publications that are documenting various al-
> gorithms, and also in planning of computer program development,
> for sketching out the structure of the program before the actual
> coding takes place."

The term "pseudocode" does not refer to a specific set of rules for ex-
pressing and formatting algorithms. Indeed, the Wikipedia article goes
on to give examples of pseudocode based on the Fortran, Pascal, and C

[3]From: Wikipedia contributors, "Pseudocode", Wikipedia, The Free Encyclopedia.
http://en.wikipedia.org/w/index.php?title=Pseudocode&oldid=564706654 (accessed
July 18, 2013). This article cites: Justin Zobel (2004). "Algorithms" in Writing for
Computer Science (second edition). Springer. ISBN 1-85233-802-4.

computer languages. The goal of pseudocode is to provide a high-level (meaning: understandable by a human) algorithm description with sufficient detail to facilitate both analysis and conversion to a computer program. A pseudocode description of an algorithm should:

- Expose the underlying algorithm;
- Hide unnecessary detail;
- Use programming constructs where appropriate, such as looping and testing; and,
- Use natural and mathematical language where appropriate.

In this book, a pseudocode based on MATLAB programming will be used, and MATLAB keywords and variables will be displayed in a special font. In particular, a loop with index ranging from 1 through N will be enclosed in the pair of statements `for k=1:N` and `end`, and a test will be enclosed with the pair `if ...` and `end`. Subscripted variables are denoted using parentheses, so that A_{ij} would be denoted `A(i,j)`. Although MATLAB statements without a trailing semicolon generally cause printing, the trailing semicolon will be omitted here. If the pseudocode is used as a template for MATLAB code, this trailing semicolon should not be forgotten.

3.3   The Gaussian Elimination Algorithm

> Algorithms are human artifacts. They belong to the world of memory and meaning, desire and design.
> — David Berlinski
> "Go ahead and faith will come to you."
> — D'Alembert.

Notice that Gaussian elimination does not use the x values in computations in any way. They are only used in the final step of back substitution to store the solution values. Thus we work with the **augmented matrix**: an $N \times N + 1$ matrix with the RHS vector in the last column

$$W_{N \times N+1} := \begin{bmatrix} a_{11} & a_{12} & \dots & a_{1n} & b_1 \\ a_{21} & a_{22} & \dots & a_{2n} & b_2 \\ \vdots & \vdots & \ddots & \vdots & \vdots \\ a_{n1} & a_{n2} & \dots & a_{nn} & b_n \end{bmatrix}.$$

Further, its backsubstitution phase does not refer to any of the zeroed out values in the matrix W. Because these are not referred to, their positions can be used to store the multipliers m_{ij}.

Evaluation of determinants using Gaussian elimination. It is known that the elementary row operation 1 multiplies $det(A)$ by the scalar, the elementary row operation 2 does not change the value of $det(A)$ and the elementary row operation 3 multiplies $det(A)$ by -1. Based on this observation, Gaussian elimination is a very efficient way to calculate determinants. If Elimination is performed and the number of row interchanges counted we then have (after W is reduced to upper triangular)

$$\det(A) = (-1)^s w_{11} \cdot w_{22} \cdot \ldots \cdot w_{nn},$$

$s =$ total number of swaps of rows and columns.

In contrast, evaluation of a determinant by cofactor expansions takes roughly $n!$ floating point operations whereas doing it using Gaussian elimination only requires $\frac{2}{3}n^3$.

We shall see that backsubstitution is much cheaper and faster to perform than elimination. Because of this, the above combination of elimination to upper triangular form followed by backsubstitution is much more efficient than complete reduction of A to the identity (so called, *Gauss-Jordan elimination*).

If the pivot entry at some step is zero we interchange the pivot row or column with a row or column below or to the right of it so that the zero structure created by previous steps is not disturbed.

Exploiting the above observations, and assuming that pivoting is not necessary, Gaussian elimination can be written in the following algorithm.

Algorithm 3.1 (Gaussian elimination without pivoting).
Given a $N \times (N + 1)$ augmented matrix W,

```
for i=1:N-1
```
   Pivoting would go here if it were required.
```
   for j=i+1:N
      if Wi,i is too small
```
         `error('divisor too small, cannot continue')`
```
      end
```
      $m = W_{ji}/W_{ii}$
```
      for k=(i+1):(N+1)
```
         $W_{jk} = W_{jk} - mW_{ik}$
```
      end
   end
end
if WN,N is too small
```

```
    error('singular!')
  end
```

Gaussian elimination has 3 nested loops. Inside each loop, roughly N (and on average $N/2$) arithmetic operations are performed. Thus, it's pretty clear that about $O(N^3)$ floating point operations are done inside Gaussian elimination for an $N \times N$ matrix.

Exercise 3.1. *Consider two so-called "magic square" matrices.*

$$A = \begin{bmatrix} 8 & 1 & 6 \\ 3 & 5 & 7 \\ 4 & 9 & 2 \end{bmatrix} \text{ and } B = \begin{bmatrix} 16 & 2 & 3 & 13 \\ 5 & 11 & 10 & 8 \\ 9 & 7 & 6 & 12 \\ 4 & 14 & 15 & 1 \end{bmatrix}.$$

Each of the rows, columns and diagonals of A sum to the same values, and similarly for B. Gaussian elimination is written above for an augmented matrix W that is $N \times N+1$. Modify it so that it can be applied to a square matrix. Then write a computer program to do Gaussian elimination on square matrices, apply it to the matrices A and B, and use the resulting reduced matrix to compute the determinants of A and B. ($\det(A) = -360$ and $\det(B) = 0$.)

The backsubstitution algorithm is below. Backsubstitution proceeds from the last equation up to the first, and MATLAB notation for this "reverse" looping is `for i=(N-1): -1:1`.

Algorithm 3.2 (Backsubstitution). *Given an N-vector* x *for storing solution values, perform the following:*

```
x(N)=W(N,N+1)/W(N,N)
for i=(N-1):-1:1
    Compute the sum s = ∑ⱼ₌ᵢ₊₁ᴺ Wᵢ,ⱼxⱼ
    x(i)=(W(i,N+1)-s)/W(i,i)
end
```

The sum $s = \sum_{j=i+1}^{N} W_{i,j}x_j$ can be accumulated using a loop, a standard programming approach to computing a sum is the following algorithm.

Algorithm 3.3 (Accumulating the sum $\sum_{j=i+1}^{n} W_{i,j} x_j$).

```
s=0
for j=(i+1):N
  s=s+W(i,j)*x(j)
end
```

Thus the backsubstitution is given as:

Algorithm 3.4 (Backsubstitution-more detail).

```
x(N)=W(N,N+1)/W(N,N)
for i=(N-1):-1:1
   Next, accumulate ∑ⁿⱼ₌ᵢ₊₁ Wᵢ,ⱼxⱼ
   s=0
   for j=(i+1):N
     s=s+W(i,j)*x(j)
   end
   x(i)=(W(i,N+1)-s)/W(i,i)
end
```

Backsubstitution has two nested loops. The innermost loop contains one add and one multiply, for two operations, there are roughly $N(N-1)/2$ passes through this innermost loop. Thus, it is clear that, in the whole, $O(N^2)$ floating point operations are done inside backsubstitution for an $N \times N$ matrix.

Exercise 3.2. *Show that the complexity of computing $det(A)$ for an $N \times N$ matrix by repeated expansion by cofactors is at least $N!$.*

3.3.1 *Computational Complexity and Gaussian Elimination*

> "In mathematics, you don't understand things. You just get used to them".
>
> — J. von Neumann (1903–1957), quoted in: G. Zukov, *The dancing Wu Li masters*, 1979.

Computers perform several types of operations:

- *Additions of real numbers*
- *Subtraction of a real numbers*
- *Multiplication of real numbers*
- *Division of real numbers*
- *Arithmetic of integers*
- *Tests (such as "Test if the number $X > 0$?")*
- *Other logical tests*
- *Accessing memory to find a number to operate upon*
- *"Loops" meaning operations of the above type performed repeatedly until some condition is met.*

The cost (in time to execute) of each of these is highly computer dependent. Traditionally arithmetic operations on real numbers have been considered to take the most time. Memory access actually takes much more time than arithmetic and there are elaborate programming strategies to minimize the effect of memory access time. Since each arithmetic operation generally requires some memory access, numerical analysts traditionally have rolled an average memory access time into the time for the arithmetic for the purpose of estimating run time. Thus one way to estimate run time is to count the number of floating point operations performed (or even just the number of multiply's and divides). This is commonly called a "FLOP count" for FLoating point OPeration count. More elegantly it is called "estimating computational complexity". Counting floating point operations gives

- Backsubstitution for an $N \times N$ linear system takes $N(N-1)/2$ multiplies and $N(N-1)/2$ adds. This is often summarized as N^2 *FLOPS*, dropping the lower order terms.
- Gaussian elimination for an $N \times N$ linear system takes $(N^3 - N)/3$ multiplies $(N^3 - N)/3$ adds and $N(N-1)/2$ divides. This is often summarized as $\frac{2}{3}N^3$ FLOPS, dropping the lower order terms.

As an example, for a 1000×1000 linear system, Gaussian elimination takes about 1000 times as long as backsubstitution. Doubling the size to 2000×2000 requires 8 times as long to run (as $(2N)^3 = 8 \cdot (N^3)$).

Exercise 3.3. *Estimate the computational complexity of computing a dot product of two N-vectors.*

Exercise 3.4. *Estimate the computational complexity of computing a residual and then the norm of a residual.*

Exercise 3.5. *Verify the claimed FLOP count for Gaussian elimination and back substitution.*

3.4 Pivoting Strategies

"Perhaps the history of the errors of mankind, all things considered, is more valuable and interesting than that of their discoveries. Truth is uniform and narrow; it constantly exists, and does not seem to require so much an active energy as a passive aptitude of the soul in order to encounter it. But error is endlessly diversified; it has no reality, but it is the pure and simple creation of the mind that invents it."

— Benjamin Franklin,

Report of Dr. B. Franklin and other commissioners, Charged by the King of France with the examination of Animal Magnetism, as now practiced in Paris, 1784.

Gaussian elimination performs the operations

$$W_{jk} = W_{jk} - \frac{W_{ji}W_{ik}}{W_{ii}}$$

many times. This can cause serious roundoff error by division by small numbers and subtraction of near equals. Pivoting strategies are how this roundoff is minimized and its cascade through subsequent calculations controlled.

We introduce the topic of pivoting strategies with an example (likely due to Wilkinson) with exact solution $(10, 1)$

$$0.0003x + 1.566y = 1.569$$
$$0.3454x - 2.436y = 1.018 \qquad (3.3)$$
$$Solution = (10, 1)$$

In 4 significant digit base 10 arithmetic we calculate:

$$m = 0.3454/0.0003 = 1151$$

and solving first for y, we find:

$$0.0003x + 1.566y = 1.569$$
$$-1802y = -1805$$

so that $y = 1.001$. Solving further for x gives

$$.0003x = 1.569 - (1.566 \cdot 1.001) = 1.569 - 1.568 = .001$$

so that the solution is

$$(x, y) = (3.333, 1.001).$$

This is very far from the exact solution and it seems likely that the error is due to dividing by a small number in backsolving for x after getting the approximate value of 1.001 for y. We consider two strategies for overcoming this: rescaling before division (which FAILS) and swapping rows (which works).

Attempt: (Rescaling FAILS)

Multiply equation (3.3) by 1000. This gives

$$0.3000x + 1566y = 1569$$
$$0.3454x - 2.436y = 1.018.$$

We find

$$m = 0.3454/0.3000 = 1.151$$

but, however it again fails:

$$(x, y) = (3.333, 1.001).$$

Again, the failure occurs during backsubstitution in the step

$$x = [1569 - 1566y]/0.3000$$

because the divisor is small with respect to both numerators.

Attempt: (Swapping rows SUCCEEDS)

$$0.3454x - 2.436y = 1.018$$
$$0.0003x + 1.566y = 1.569.$$

We find $m = 8.686 \cdot 10^{-4}$ and, again $y = 1.001$. This time, using the first equation for the backsolve yields $x = 10.00$, a much better approximation.

This example suggests that pivoting, meaning to swap rows or columns, is the correct approach. The choice of which rows or columns to swap is known as a pivoting strategy. Common ones include:

Mathematical partial pivoting: Interchange only the rows, not the columns, when the pivot entry $W_{ii} = 0$. This strategy *is not sufficient* to eliminate roundoff errors. Even if pivoting is done when $W_{ii} = 0$ *to numerical precision*, this strategy is not sufficient.

Simple partial pivoting: Interchange rows to maximize $|W_{ji}|$ over $j \geq i$.

Algorithm 3.5 (Simple partial pivoting). *Given a column* i

Find row $j \geq i$ so that $|W_{ji}|$ is maximized.
Swap rows j and i.

Simple partial pivoting is a common strategy, but there are better ones.
Scaled partial pivoting: Interchange only rows so that the pivot entry W_{ii} is the element in column i on or below the diagonal which is largest relative to the size of the whole row that entry is in.

Algorithm 3.6 (Scaled partial pivoting). *Given a column* i

(1) Compute $d_j := \max_{i \leq k \leq N} |W_{jk}|$.
(2) Find row $j \geq i$ so that $|W_{ji}|/d_i$ is maximized.
(3) Swap rows i and j.

The following refinement of Algorithm 3.6 breaks the steps of that algorithm into detailed pseudocode. The pseudocode in this algorithm is intended to stand alone so that it can be "called" by name from another, larger algorithm. Separate groups of code of this nature are often called "functions", "subroutines", or "procedures". The MATLAB syntax for a function is:

function [*"return" values*] = *function name*(*arguments*)

There may be zero or more return values and zero or more arguments. If there are zero or one return values, the brackets ("[" and "]") can be omitted.

Algorithm 3.7 (Scaled partial pivoting (detailed)).
Given a row number, i, an $N \times N$ matrix W, and the value N, return the row number pivotrow *with which it should be swapped.*

```
function pivotrow = scaled_partial_pivoting(i,W,N)
% First, find the maximum in each row.
for j=i:N
  d(j) = abs(W(j,j))
  for k=i+1:N
    if d(j) < abs(W(j,k))
      d(j) = abs(W(j,k))
    end
```

```
    end
end

% Second, find the pivot row
pivotrow=i
pivot = abs(W(i,i))/d(i)
for j=i+1:N
  if pivot < abs(W(j,i))/d(j)
    pivot = abs(W(j,i))/d(j)
    pivotrow = j
  end
end
```

Scaled partial pivoting is a very commonly used strategy. It gives a good balance between stability and computational cost.

Exercise 3.6. *Give a detailed elaboration of the partial pivoting algorithm (at a similar level of detail as the scaled partial pivoting algorithm).*

Exercise 3.7. *Multiply the first equation in (3.3) by 10,000. Show that Scaled partial pivoting yields the correct answer in four-digit arithmetic, but partial pivoting does not.*

Full pivoting: Interchange rows and columns to maximize $|W_{ik}|$ over $i \geq j$ and $k \geq j$. Full pivoting is less common because interchanging columns reorders the solution variables. The reordering must be stored as an extra $N \times N$ matrix to recover the solution in the correct variables after the process is over.

Example 3.1. Suppose at one step of Gaussian elimination, the augmented system is

$$W = \begin{bmatrix} 1.0 & 2.0 & 3.0 & 4.0 & 1.7 \\ 0 & 10^{-10} & 2.0 & 3.0 & 6.0 \\ 0 & 2.0 & 3.0 & 1.0 & -1000.0 \\ 0 & 3.0 & 2.0 & -5.0 & 35.0 \end{bmatrix}.$$

The RHS vector and active submatrix are partitioned with lines (that are not stored in W). The pivot entry is now the $2,2$ entry (currently $W(2,2) = 10^{-10}$). For the different pivoting strategies we would have

- **Mathematical pivoting:** no swapping since $10^{-10} \neq 0$.

- **Partial pivoting:** Row 2 swap with Row 4 since 3.0 is the largest entry below 10^{-10}.
- **Scaled partial pivoting:** Row 2 swap with Row 3 since $2.0/3.0 > 3.0/5.0$.
- **Full pivoting:** Row 2 swap with Row 4 and Column 2 swap with column 4 since -5.0 is the largest entry in absolute value in the active submatrix.

Putting scaled partial pivoting into the Gaussian Elimination Algorithm 3.1 yields the following algorithm. In this algorithm, a vector, p, is also computed to keep track of row interchanges, although it is not needed when applying Gaussian Elimination to an augmented matrix. This algorithm is written so that it can be applied to any square or rectangular $N \times M$ matrix with $M \geq N$.

Algorithm 3.8 (Gaussian elimination with scaled partial pivoting).
Given a $N \times M$ $(M \geq N)$ matrix W,

```
for i = 1:N
  p(i) = i
end
for i = 1:(N-1)
  j = scaled_partial_pivoting(i,W,N)
  Interchange p(i) and p(j)
  Interchange rows i and j of W
  for j=(i+1):N
    m = W(j,i)/W(i,i)
    for k =(i+1):M
      W(j,k) = W(j,k) - m*W(i,k)
    end
  end
end
if W_{N,N} is too small
  error('Matrix is singular!')
end
```

Interchanging two components of p is accomplished by:

Algorithm 3.9 (Interchange components of p). *Given a vector p and two indices i and j.*

```
temporary = p(i)
p(i) = p(j)
p(j) = temporary
```

Interchanging two rows of W is similar, but requires a loop.

Exercise 3.8. *Write detailed pseudocode for interchanging two rows of W.*

Exercise 3.9. *Solve the 3×3 linear system with augmented matrix given below by hand executing the Gaussian elimination with scaled partial pivoting algorithm:*

$$W = \begin{bmatrix} -1 & 2 & -1 & 0 \\ 0 & -1 & 2 & 1 \\ 2 & -1 & 0 & 0 \end{bmatrix}.$$

Exercise 3.10. *Suppose you are performing Gaussian elimination on a square matrix A. Suppose that in your search for a pivot for column i using simple partial pivoting you discover that $\max_{j \geq i} |W(j,i)|$ is exactly zero. Show that the matrix A must be singular. Would the same fact be true if you were using scaled partial pivoting?*

Exercise 3.11. *In Algorithm 3.8, the matrix W is an $N \times M$ matrix and it employs Algorithm 3.6. In that algorithm d_j is constructed for $i \leq k \leq N$ and not $i \leq k \leq M$. When $M > N$, explain why it is a reasonable to ignore some columns of W when pivoting.*

3.5 Tridiagonal and Banded Matrices

> "The longer I live, the more I read, the more patiently I think,
> and the more anxiously I inquire, the less I seem to know."
> — John Adams

Gaussian elimination is much faster for banded matrices (in general) and especially so for tridiagonal ones (in particular).

Definition 3.1. An $N \times N$ matrix A is tridiagonal if $A_{ij} = 0$ for $|i-j| > 1$.

Thus a tridiagonal matrix is one that takes the form:

$$
\begin{bmatrix}
d_1 & c_1 & 0 & 0 & \cdots & 0 \\
a_2 & d_2 & c_2 & 0 & \cdots & 0 \\
0 & a_3 & d_3 & c_3 & \cdots & 0 \\
& \ddots & \ddots & \ddots & & \\
0 & \cdots & 0 & a_{N-1} & d_{N-1} & c_{N-1} \\
0 & \cdots & 0 & 0 & a_N & d_N
\end{bmatrix}
\begin{bmatrix}
x_1 \\ x_2 \\ x_3 \\ \vdots \\ x_{N-1} \\ x_N
\end{bmatrix}
=
\begin{bmatrix}
b_1 \\ b_2 \\ b_3 \\ \vdots \\ b_{N-1} \\ b_N
\end{bmatrix}.
$$

Performing elimination without pivoting does not alter the tridiagonal structure. Thus the zeroes need not be stored (saving lots of storage: from $O(N^2)$ to $O(N)$). There is no point in doing arithmetic on those zeroes, either, reducing FLOPS from $O(N^3)$ to $O(N)$. There are two common ways to store a tridiagonal linear system.

Method 1: storage as 4 vectors by:

$$
\vec{a} := (0, a_2, a_3, \cdots, a_{N-1}, a_N)
$$
$$
\vec{d} := (d_1, d_2, d_3, \cdots, d_{N-1}, d_N)
$$
$$
\vec{c} := (c_1, c_2, c_3, \cdots, c_{N-1}, 0)
$$
$$
\vec{b} := (b_1, b_2, b_3, \cdots, b_{N-1}, b_N).
$$

Stored in this form the elimination and backsubstitution algorithms are as follows.

Algorithm 3.10 (Tridiagonal Elimination). *Given 4 N-vectors a, d, c, b, satisfying $a(1) = 0.0$ and $c(N) = 0.0$*

```
for i = 2:N
  if d(i-1) is zero
    error('the matrix is singular or pivoting is required')
  end
  m = a(i)/d(i-1)
  d(i) = d(i) - m*c(i-1)
  b(i) = b(i) - m*b(i-1)
end
if d(N) is zero
  error('the matrix is singular')
end
```

Clearly, tridiagonal Gaussian elimination has one loop. Inside the loop, roughly five arithmetic operations are performed. Thus, it is clear that, on the whole, $O(N)$ floating point operations (more precisely 5N-5 FLOPS) are done inside tridiagonal Gaussian elimination for an $N \times N$ matrix.

The backsubstitution algorithm is as follows.[4]

Algorithm 3.11 (Tridiagonal Backsubstitution). *Given an extra N-vector x to store the solution values, perform the following:*

```
x(N) = b(N)/d(N)
for i = N-1:-1:1
  x(i)=( b(i) - c(i)*x(i+1) )/d(i)
end
```

Example 3.2. A tridiagonal matrix that frequently occurs is `tridiag(-1,2,-1)` or

$$
\begin{bmatrix}
2 & -1 & 0 & 0 & \ldots & 0 \\
-1 & 2 & -1 & 0 & \ldots & 0 \\
0 & -1 & 2 & -1 & \ldots & 0 \\
& & \ddots & \ddots & \ddots & \\
0 & \ldots & 0 & -1 & 2 & -1 \\
0 & \ldots & 0 & 0 & -1 & 2
\end{bmatrix}.
$$

The first step in GE for this matrix is to replace:

$Row2 <= Row2 - (-1/2)Row1.$

This gives

$$
\begin{bmatrix}
2 & -1 & 0 & 0 & \ldots & 0 \\
0 & 1.5 & -1 & 0 & \ldots & 0 \\
0 & -1 & 2 & -1 & \ldots & 0 \\
& & \ddots & \ddots & \ddots & \\
0 & \ldots & 0 & -1 & 2 & -1 \\
0 & \ldots & 0 & 0 & -1 & 2
\end{bmatrix}.
$$

This zeroes out the entire first column.

Backsubstitution for tridiagonal matrices is also an $O(N)$ algorithm since there is one loop with a subtraction, a multiplication, and a division.

In summary, tridiagonal system solution without pivoting requires:

- $5N - 5$ adds, multiplies and divides for elimination, and,
- $3N - 2$ adds, multiplies and divides for backsubstitution.

---

[4]Recall that the syntax "`for i=N-1:-1:1`" means that the loop starts at `i=N-1` and `i` decreases to 1.

More generally, for a banded, sparse matrix with half bandwidth p (and thus full bandwidth $2p+1$) banded sparse Gaussian elimination takes $O(p^2 N)$ FLOPS.

Method 2: Storage as a banded matrix with bandwidth three or half bandwidth $p = 1$.

In this case we store the augmented matrix as

$$W_{4 \times N} := \begin{bmatrix} b_1 & b_2 & \dots & b_N \\ c_1 & c_2 & \dots & 0 \\ d_1 & d_2 & \dots & d_N \\ 0 & a_2 & \dots & a_N \end{bmatrix}.$$

Modification of the Gaussian elimination algorithm for this alternative storage method is given below.

Algorithm 3.12 (Tridiagonal Gaussian Elimination: Band Storage).
Given a $4 \times n$ augmented matrix W, with $W_{41} = 0.0$ and $W_{2N} = 0.0$.

```
for i = 2:N
  if W(3,i-1) is zero
    error('the matrix is singular or pivoting is required')
  end
  m = W(4,i)/W(3,i-1)
  W(3,i) = W(3,i) - m*W(2,i-1)
  W(1,i) = W(1,i) - m*W(1,i-1)
end
if W(3,N) is zero
  error('the matrix is singular.')
end
```

Exercise 3.12. *Give a pseudocode algorithm for backsubstitution for tridiagonal matrices stored in band form.*

Exercise 3.13. *Extend the algorithms given here to general banded systems with half bandwidth $p < N/2$.*

Exercise 3.14. *What is the operation count for an $N \times N$ system for Gaussian elimination? Back substitution? If the matrix is tridiagonal, then what are the operation counts?*

Exercise 3.15. *If a 10000×10000 tridiagonal linear system takes 2 minutes to solve using tridiagonal elimination plus backsubstitution, estimate how* long it would take to solve using the full *GE* plus full backsubstitution algorithms. *(This explains why it makes sense to look at the special case of tridiagonal matrices.)*

Exercise 3.16. *Verify that Gaussian elimination requires* $O(p^2 N)$ *FLOPS for an* $N \times N$ *banded matrix with half-bandwidth* p.

Exercise 3.17. *The above algorithms for tridiagonal Gaussian elimination contain lines such as* "if W(3,N) is zero" *or* "if d(N) is zero". *If you were writing code for a computer program, how would you interpret these lines?*

Exercise 3.18. *Write a computer program to solve the tridiagonal system* $A = tridiag(-1, 2, -1)$ *using tridiagonal Gaussian elimination with band storage, Algorithm 3.12. Test your work by choosing the solution vector of all* $1's$ *and RHS containing the row sums of the matrix. Test your work for system sizes* $N = 3$, *and* $N = 1000$.

3.6   The *LU* Decomposition

"Measure what is measurable, and make measurable what is not so."

— Galilei, Galileo (1564–1642), Quoted in H. Weyl "Mathematics and the Laws of Nature" in I Gordon and S. Sorkin (eds.) The Armchair Science Reader, New York: Simon and Schuster, 1959.

"Vakmanschap is meesterschap."

— (Motto of Royal Grolsch NV, brewery.)

Suppose we could factor the $N \times N$ matrix A as the product

$$A = LU, \qquad L : \text{lower triangular} , \quad U : \quad \text{upper triangular}.$$

Then, we can solve the linear system $Ax = b$ without Gaussian elimination in two steps:

(1) Forward substitution: solve $Ly = b$.
(2) Backward substitution: solve $Ux = y$.

Step 1. Forward solve for y

$$Ly = b$$

$$\Leftrightarrow$$

$$\begin{bmatrix} \ell_{11} & 0 & 0 & \dots 0 & 0 \\ \ell_{21} & \ell_{22} & 0 & \dots 0 & 0 \\ \vdots & \vdots & \ddots & \vdots & \ddots & \vdots \\ \ell_{N-1,1} & \ell_{N-1,2} & \ell_{N-1,3} & \dots \ell_{N-1,N-1} & 0 \\ \ell_{N1} & \ell_{N2} & \ell_{N,3} & \dots \ell_{N,N-1} & \ell_{N,N} \end{bmatrix} \begin{bmatrix} y_1 \\ y_2 \\ \vdots \\ y_{N-1} \\ y_N \end{bmatrix} = \begin{bmatrix} b_1 \\ b_2 \\ \vdots \\ b_{N-1} \\ b_{NN} \end{bmatrix}$$

so

$$\ell_{11} y_1 = b_1 \Rightarrow y_1 = b_1/\ell_{11},$$
$$\ell_{21} y_1 + \ell_{22} y_2 = b_2 \quad \Rightarrow \quad y_2 = (b_2 - \ell_{21} y_1)/\ell_{22},$$

and so on.

Step 2. Backward solve $Ux = y$ for x

$$Ux = y$$

$$\Leftrightarrow$$

$$\begin{bmatrix} u_{11} & u_{12} & u_{1,3} & \dots u_{1,N-1} & u_{1N} \\ 0 & u_{22} & u_{23} & \dots u_{2,N-1} & u_{2N} \\ \vdots & \vdots & \ddots & \vdots & \ddots & \vdots \\ 0 & 0 & 0 & \dots u_{N-1,N-1} & u_{N,N} \\ 0 & 0 & 0 & \dots 0 & u_{N,N} \end{bmatrix} \begin{bmatrix} x_1 \\ x_2 \\ \vdots \\ x_{N-1} \\ x_N \end{bmatrix} = \begin{bmatrix} y_1 \\ y_2 \\ \vdots \\ y_{N-1} \\ y_{NN} \end{bmatrix},$$

so

$$u_{NN} x_N = y_N \quad \Rightarrow \quad x_N = y_N/u_{NN}$$

and

$$u_{N-1,N-1} x_{N-1} + u_{N-1,N} x_N = y_{N-1} \quad \Rightarrow$$
$$x_{N-1} = (y_{N-1} - u_{N-1,N} x_N)/u_{N-1,N-1}.$$

Thus, once we compute a factorization $A = LU$ we can solve linear systems relatively cheaply. This is especially important if we must solve many linear systems with the same $A = LU$ and different RHS's b. First consider the case without pivoting.

Theorem 3.1 (Remarkable Algorithmic Fact). *If no pivoting is used and the Gaussian elimination algorithm stores the multipliers m_{ij} below the* *diagonal of A the algorithm computes the LU factorization of A where L and U are given by*

$$L = \begin{bmatrix} 1 & 0 & 0 & \dots 0 & 0 \\ m_{21} & 1 & 0 & \dots 0 & 0 \\ \vdots & \vdots & \ddots & \vdots & \vdots & \vdots \\ m_{N-1,1} & m_{N-1,2} & m_{N-1,3} & \dots 1 & 0 \\ m_{N1} & m_{N2} & m_{N,3} & \dots m_{N,N-1} & 1 \end{bmatrix},$$

$$U = \begin{bmatrix} u_{11} & u_{12} & u_{1,3} & \dots u_{1,N-1} & u_{1N} \\ 0 & u_{22} & u_{23} & \dots u_{2,N-1} & u_{2N} \\ \vdots & \vdots & \ddots & \vdots & \ddots & \vdots \\ 0 & 0 & 0 & \dots u_{N-1,N-1} & u_{N,N} \\ 0 & 0 & 0 & \dots 0 & u_{N,N} \end{bmatrix}.$$

Exercise 3.19. *Prove Theorem 3.1 for the 3×3 case using the following steps.*

(1) *Starting with a 3×3 matrix A, perform one column of row reductions, resulting in a matrix*

$$\overline{L}_1 = \begin{bmatrix} 1 & 0 & 0 \\ -m_{21} & 1 & 0 \\ -m_{31} & 0 & 1 \end{bmatrix}$$

with $-m_{n,1}$ for $n = 2,3$ denoting the multipliers used in the row reduction.

(2) *Consider the matrix*

$$L_1 = \begin{bmatrix} 1 & 0 & 0 \\ m_{21} & 1 & 0 \\ m_{31} & 0 & 1 \end{bmatrix}$$

and show that

(a) *$\overline{L}_1 A = U_1$, and*
(b) *$\overline{L}_1 L_1 = I$, where I is the identity matrix, so $\overline{L}_1 = (L_1)^{-1}$.*

Hence, $A = L_1 U_1$.

(3) *Similarly, perform one column of row reductions on the second column of U_1 and construct the matrix*

$$\overline{L}_1 = \begin{bmatrix} 1 & 0 & 0 \\ 0 & 1 & 0 \\ 0 & -m_{32} & 1 \end{bmatrix}.$$

(4) Show that

(a) $\overline{L}_2 U_1 = U_2$, and

(b) $\overline{L}_2 L_2 = I$, where I is the identity matrix, so $\overline{L}_2 = (L_2)^{-1}$, and

(c) *(this is the surprising part)*

$$L_1 L_2 = \begin{bmatrix} 1 & 0 & 0 \\ m_{21} & 1 & 0 \\ m_{31} & m_{32} & 1 \end{bmatrix}$$

so that $A = L_1 L_2 U_2$.

Exercise 3.20. *Prove Theorem 3.1 for the general case, using Exercise 3.19 as a model.*

Remark 3.1. When solving systems with multiple RHS's, it is common to compute L and U in double precision and store in the precision sought in the answer (either single or double). This gives extra accuracy without extra storage. Precisions beyond double are expensive, however, and are used sparingly.

Remark 3.2. Implementations of Gaussian Elimination combine the two matrices L and U together instead of storing the ones on the diagonal of L and all the zeros of L and U, a savings of storage for N^2 real numbers. The combined matrix is

$$W = \begin{bmatrix} u_{11} & u_{12} & u_{1,3} & \cdots u_{1,N-1} & u_{1N} \\ m_{21} & u_{22} & u_{23} & \cdots u_{2,N-1} & u_{2N} \\ \vdots & \vdots & \ddots & \vdots & \ddots & \vdots \\ m_{N-1,1} & m_{N-1,2} & m_{N-1,3} & \cdots u_{N-1,N-1} & u_{N,N} \\ m_{N,1} & m_{N,2} & m_{N,3} & \cdots m_{N,N-1} & u_{N,N} \end{bmatrix}.$$

Example 3.3. Suppose A is the 4×4 matrix below.

$$\begin{bmatrix} 3 & 1 & -2 & -1 \\ 2 & -2 & 2 & 3 \\ 1 & 5 & -4 & -1 \\ 3 & 1 & 2 & 3 \end{bmatrix}.$$

Performing Gauss elimination without pivoting (exactly as in the algorithm) and storing the multipliers gives

$$W = \begin{bmatrix} 3 & 1 & -2 & -1 \\ \frac{2}{3} & -\frac{8}{3} & \frac{10}{3} & \frac{11}{3} \\ \frac{1}{3} & -\frac{7}{4} & \frac{5}{2} & \frac{23}{4} \\ 1 & 0 & \frac{8}{5} & -\frac{26}{5} \end{bmatrix}.$$

Thus, $A = LU$ where

$$L = \begin{bmatrix} 1 & 0 & 0 & 0 \\ \frac{2}{3} & 1 & 0 & 0 \\ \frac{1}{3} & -\frac{7}{4} & 1 & 0 \\ 1 & 0 & \frac{8}{5} & 1 \end{bmatrix} \text{ and } U = \begin{bmatrix} 3 & 1 & -2 & -1 \\ 0 & -\frac{8}{3} & \frac{10}{3} & \frac{11}{3} \\ 0 & 0 & \frac{5}{2} & \frac{23}{4} \\ 0 & 0 & 0 & -\frac{26}{5} \end{bmatrix}.$$

Exercise 3.21. *Algorithm 3.12 describes tridiagonal Gaussian elimination. Modify that algorithm to store the multipliers in the matrix W so that it computes both the lower and upper tridiagonal factors.*

Exercise 3.22. *Suppose A has been factored as $A = LU$ with L and U given below. Use this factorization to solve $Ax = e_3$, where $e_3 = (0, 0, 1)^t$.*

$$L = \begin{bmatrix} 1 & 0 & 0 \\ 2 & 1 & 0 \\ 3 & 4 & 1 \end{bmatrix}, U = \begin{bmatrix} 2 & 3 & 4 \\ 0 & 1 & 0 \\ 0 & 0 & 3 \end{bmatrix}.$$

Exercise 3.23. *Algorithm 3.8 describes the algorithm for Gaussian elimination with scaled partial pivoting for an augmented matrix W, but it does not employ the combined matrix factor storage described in Remark 3.2. Modify Algorithm 3.8 so that*

(1) It applies to square matrices; and,
(2) It employs combined matrix factor storage.

The question remains: What happens when pivoting is required? To help answer this question, we need to introduce the concept of a "permutation" to make the notion of swapping rows clear.

Definition 3.2. A permutation *vector* is a rearrangement of the vector

$$\vec{p} = [1, 2, 3, \cdots, N]^t.$$

A permutation *matrix* is an $N \times N$ matrix whose columns are rearrangements of the columns of the $N \times N$ identity matrix. This means that there is a permutation vector such that

$$j^{th} \text{ column of } P = p(j)^{th} \text{ column of } I.$$

For example, if $N = 2$, the permutation matrices are

$$P_1 = \begin{bmatrix} 0 & 1 \\ 1 & 0 \end{bmatrix} \text{ and } P_2 = \begin{bmatrix} 1 & 0 \\ 0 & 1 \end{bmatrix}.$$

Note that

$$P_1 \begin{bmatrix} x_1 \\ x_2 \end{bmatrix} = \begin{bmatrix} x_2 \\ x_1 \end{bmatrix} \text{ and } P_1^{-1} \begin{bmatrix} x_1 \\ x_2 \end{bmatrix} = \begin{bmatrix} x_2 \\ x_1 \end{bmatrix}.$$

If $\overrightarrow{p} = (2,1)$ then we compute $\overrightarrow{y} = P^{-1}\overrightarrow{x}$ by

```
for i = 1:2
  y(i) = x( p(i) )
end
```

More generally, we compute $\overrightarrow{y} = P^{-1}\overrightarrow{x}$ by

```
for i=1:N
  y(i)=x( p(i) )
end
```

Theorem 3.2 ($A = PLU$ **factorization**). *Gaussian elimination with partial pivoting, as presented in Algorithm 3.8 and modified in Exercise 3.23, computes the permutation vector, $\overrightarrow{p}$, as part of the elimination process and stores both the multipliers and the upper triangular matrix in the combined matrix W. Thus, it constructs the factorization*

$$A = PLU,$$

where P is the permutation matrix corresponding to the vector $\overrightarrow{p}$.

Proof. The essential part of the proof can be seen in the 3×3 case, so that case will be presented here.

The first step in Algorithm 3.8 is to find a pivot for the first column. Call this pivot matrix P_1. Then row reduction is carried out for the first column, with the result

$$A = (P_1^{-1}P_1)A = P_1^{-1}(P_1 A) = P_1^{-1}L_1 U_1,$$

where

$$L_1 = \begin{bmatrix} 1 & 0 & 0 \\ m_{21} & 1 & 0 \\ m_{31} & 0 & 1 \end{bmatrix} \text{ and } U_1 = \begin{bmatrix} u_{11} & * & * \\ 0 & * & * \\ 0 & * & * \end{bmatrix}$$

where the asterisks indicate entries that might be non-zero.

The next step is to pivot the second column of U_1 from the diagonal down, and then use row-reduction to factor it:

$$A = P_1^{-1}L_1(P_2^{-1}L_2 U)$$

where

$$L_2 = \begin{bmatrix} 1 & 0 & 0 \\ 0 & 1 & 0 \\ 0 & m_{32} & 1 \end{bmatrix} \text{ and } U = \begin{bmatrix} u_{11} & u_{12} & u_{13} \\ 0 & u_{22} & u_{23} \\ 0 & 0 & u_{33} \end{bmatrix}.$$

Finally, it must be shown that $L_1 P_2^{-1} L_2$ can be expressed as $\tilde{P}^{-1}L$. It is clear that

$$L_1 P_2^{-1} L_2 = (P_2^{-1} P_2) L_1 P_2^{-1} L_2 = P_2^{-1}(P_2 L_1 P_2^{-1}) L_2.$$

There are only two possibilities for the permutation matrix P_2. It can be the identity, or it can be

$$P_2 = \begin{bmatrix} 1 & 0 & 0 \\ 0 & 0 & 1 \\ 0 & 1 & 0 \end{bmatrix}.$$

If P_2 is the identity, then $P_2^{-1}(P_2 L_1 P_2^{-1})L_2 = L_1 L_2$ and is easily seen to be lower triangular. If not,

$$P_2 L_1 P_2^{-1} = \begin{bmatrix} 1 & 0 & 0 \\ 0 & 0 & 1 \\ 0 & 1 & 0 \end{bmatrix} \begin{bmatrix} 1 & 0 & 0 \\ m_{21} & 1 & 0 \\ m_{31} & 0 & 1 \end{bmatrix} \begin{bmatrix} 1 & 0 & 0 \\ 0 & 0 & 1 \\ 0 & 1 & 0 \end{bmatrix}$$

$$= \begin{bmatrix} 1 & 0 & 0 \\ m_{31} & 1 & 0 \\ m_{21} & 0 & 1 \end{bmatrix}.$$

Hence,

$$P_2 L_1 P_2^{-1} L_2 = \begin{bmatrix} 1 & 0 & 0 \\ m_{31} & 1 & 0 \\ m_{21} & m_{32} & 1 \end{bmatrix},$$

a lower triangular matrix. □

Exercise 3.24. *Complete the proof of Theorem 3.2 for $N \times N$ matrices.* **Hint:** *It is important to realize that the fact that $P_2 L_1 P_2^{-1} L_2$ turns out to be lower triangular depends strongly on the permutation matrix P_2 involving indices greater than the indices of columns of L_1 with non-zeros below the diagonal.*

Given the factorization $A = PLU$, the solution of $Ax = b$ is then found in three steps.

Algorithm 3.13 (Solving $PLUx = b$).

(1) Compute $\vec{d} = P^{-1}\vec{b}$, i.e., rearrange $\vec{b}$ by:

```
for k=1:N
  d(k)=b(p(k))
end
```

(2) Forward solve $L\vec{y} = \vec{d}$.
(3) Backsolve $U\vec{x} = \vec{y}$.

Example 3.4. Suppose in solving a 3×3, elimination swaps rows 1 and row 2. Then $p = (1, 2, 3)$ is changed to $p = (2, 1, 3)$ at the end of elimination. Let $b = (1, 3, 7)^t$, $p = (2, 1, 3)^t$. Then $d = P^{-1}b = (3, 1, 7)^t$.

Remark 3.3.

(1) Factoring $A = PLU$ takes $O(N^3)$ FLOPS but using it thereafter for backsolves only takes $O(N^2)$ FLOPS.
(2) If A is symmetric and positive definite then the PLU decomposition can be further refined into an LL^t decomposition known as the Cholesky decomposition. It takes about $1/2$ the work and storage of PLU. Gaussian elimination for SPD matrices do not require pivoting, an important savings in time and storage.

Exercise 3.25. *Find the LU decomposition of*
$$\begin{bmatrix} 3 & 9 \\ 2 & 7 \end{bmatrix}.$$

Exercise 3.26. *Given the LU decomposition*
$$\begin{bmatrix} 2 & 3 \\ 8 & 11 \end{bmatrix} = \begin{bmatrix} 1 & 0 \\ 4 & 1 \end{bmatrix} \begin{bmatrix} 2 & 3 \\ 0 & -1 \end{bmatrix}$$
use it to solve the linear system
$$2x + 3y = 0$$
$$8x + 11y = 1.$$

Exercise 3.27. *Write a computer program to perform Gaussian elimination on a square matrix, A, using partial pivoting (Algorithm 3.8 as modified in Exercise 3.23).*

(1) *At the end of the algorithm, reconstruct the matrices P, L, and U and compute a relative norm $\|A - PLU\|/\|A\|$. (You can use the Frobenius norm $\|A\|_{fro}^2 = \sum_{ij} |A_{ij}|^2$.) The norm should be zero or nearly zero. (Alternatively, perform the calculation $\|A - PLU\|/\|A\|$ without explicitly constructing P, L, and U.)*

(2) *To help debug your work, at the beginning of each of the column reduction steps (the second* for i= *loop), reconstruct the matrices P, L, and U and compute a norm $\|A - PLU\|$. The norm should be zero or nearly zero each time through the loop. (Alternatively, compute the norm without reconstructing the matrix factors.) Once you are confident your code is correct, you can eliminate this debugging code.*

(3) *Test your code on the $N \times N$ matrix consisting of all ones everywhere except that the diagonal values are zero. Use several different values of N as tests. The 3×3 case looks like*

$$A = \begin{bmatrix} 0 & 1 & 1 \\ 1 & 0 & 1 \\ 1 & 1 & 0 \end{bmatrix}. \tag{3.4}$$

Exercise 3.28. *(This exercise continues Exercise 3.27.) Write a computer program to perform the backsubstitution steps, given the compressed matrix W arising from Gaussian elimination with scaled partial pivoting. Test your work by applying it to the $N \times N$ matrix A described in Exercise 3.27 with right side given by $b = (N - 1, N - 1, \ldots, N - 1)^t$, whose solution is $x = (1, 1, \ldots, 1)^t$. Use several values of N for your tests.*

Chapter 4

Norms and Error Analysis

"fallor ergo sum."
— Augustine.

In Gaussian elimination there are a large number of calculations. Each operation depends upon all previous ones. Thus, round off error occurs and propagates. It is critically important to understand and quantify precisely what "numerically singular" or "ill conditioned" means, to quantify it and to predict its effect on solution cost and accuracy. We begin this study in this chapter.

4.1 FENLA and Iterative Improvement

An expert is someone who knows some of the worst mistakes that can be made in his subject, and how to avoid them.
— Heisenberg, Werner (1901–1976), Physics and Beyond. 1971.

If the matrix is numerically singular or ill-conditioned, it can be difficult to obtain the accuracy needed in the solution of the system $Ax = b$ by Gaussian elimination alone. *Iterative improvement* is an algorithm to increase the accuracy of a solution to $Ax = b$. The basic condition needed is that $Ax = b$ can be solved to at least one significant digit of accuracy. Iterative improvement is based on the Fundamental Equation of Numerical Linear Algebra (the "FENLA").

Theorem 4.1 (FENLA). *Let $A_{N \times N}$, $b_{n \times 1}$ and let x be the true solution to $Ax = b$. Let $\hat{x}$ be some other vector. The error $e := x - \hat{x}$ and residual $r := b - A\hat{x}$ are related by*

$$Ae = r.$$

Proof. Since $e = x - \hat{x}$, $Ae = A(x - \hat{x}) = Ax - A\hat{x}$. Then, since $Ax = b$,

$$Ae = Ax - A\hat{x} = b - A\hat{x} = r.$$

$\square$

Given a candidate for a solution $\hat{x}$, if we could find its error $\hat{e}(= x - \hat{x})$, then we would recover the true solution

$$x = \hat{x} + \hat{e} \quad (\text{since } \hat{x} + \hat{e} = \hat{x} + (x - \hat{x}) = x).$$

Thus we can say the following two problems are equivalent:

Problem 1: Solve $Ax = b$.
Problem 2: Guess $\hat{x}$, compute $\hat{r} = b - A\hat{x}$, solve $A\hat{e} = \hat{r}$, and set $x = \hat{x} + \hat{e}$.

This equivalence is the basis of iterative improvement.

Algorithm 4.1 (Iterative Improvement). *Given a matrix A, a RHS vector b, and a precision t, find an approximate solution x to the equation $Ax = b$ with at least t correct significant digits.*

Compute the $A = LU$ factorization of A in working precision
Solve $Ax = b$ for candidate solution $\hat{x}$ in working precision
`for k=1:maxNumberOfIterations`
    Calculate the residual

$$r = b - A\hat{x} \tag{4.1}$$

    in extended *precision*
    Solve $Ae = r$ (by doing 2 backsolves in working precision) for
        an approximate error $\hat{e}$
    Replace $\hat{x}$ with $\hat{x} + \hat{e}$
    `if` $\|\hat{e}\| \le 10^{-t}\|\hat{x}\|$
        `return`
    `end`
`end`
`error(`'*The iteration did not achieve the required error.*'`)`

It is not critical to perform the residual calculation (4.1) in higher precision than that used to store the matrix A and vector x, but it substantially improves the algorithm's convergence and, in cases with extremely large condition numbers, is required for convergence.

Using extended precision for the residual may require several iteration steps, and the number of steps needed increases as A becomes more ill-conditioned, but in all cases, it is much cheaper than computing the LU decomposition of A itself in extended precision. Thus, it is almost always performed in good packages.

Example 4.1. Suppose the matrix A is so ill conditioned that solving with it only gives 2 significant digits of accuracy. Stepping through iterative improvement we have:

$\widehat{x} = 2$: sig-digits

Calculate $\widehat{r} = b - A\widehat{x}$

Solve $A\widehat{e} = \widehat{r}$

$\widehat{e} = 2$: sig-digits

Then $\widehat{x} \Leftarrow \widehat{x} + \widehat{e}$: 4 significant digits.

$\widehat{e} = 2$: sig-digits

Then $\widehat{x} \Leftarrow \widehat{x} + \widehat{e}$: 6 significant digits, and so on until the desired accuracy is attained.

Example 4.2. On a 3 significant digit computer, suppose we solve $Ax = b$ where (to 3 significant digits)

$$b = [5.90 \quad 7.40 \quad 10.0]^t$$

and

$$A = \begin{bmatrix} 1.00 & 1.20 & 1.50 \\ 1.20 & 1.50 & 2.00 \\ 1.50 & 2.00 & 3.00 \end{bmatrix}.$$

The exact solution of $Ax = b$ is

$$x = [2.00 \quad 2.00 \quad 1.00]^t.$$

Step 1: Computing $A = LU$ in working precision (using 3 significant digits in this example) gives

$$A = \begin{bmatrix} 1.00 & 1.20 & 1.50 \\ 1.20 & 1.50 & 2.00 \\ 1.50 & 2.00 & 3.00 \end{bmatrix} = \begin{bmatrix} 1.00 & 0.00 & 0.00 \\ 1.20 & 1.00 & 0.00 \\ 1.50 & 3.33 & 1.00 \end{bmatrix} \begin{bmatrix} 1.00 & 1.20 & 1.50 \\ 0.00 & 0.0600 & 0.200 \\ 0.00 & 0.00 & 0.0840 \end{bmatrix}.$$

Step 2: Solving $Ax = b$ in 3 significant digit arithmetic using 2 backsolves ($Ly = b$ and $Ux = y$) gives

$$\widehat{x} = [1.87 \quad 2.17 \quad 0.952]^t.$$

Step 3: A "double precision" (6 digit) calculation of the residual gives

$$r = [-0.00200 \quad -0.00300 \quad -0.00100]^t.$$

Step 4: The single precision (3 digit) solution of $LU\hat{e} = r$ is

$$\hat{e} = [0.129 \quad -0.168 \quad 0.0476]^t.$$

Step 5: Update solution

$$\hat{x} = \hat{x}_{\text{OLD}} + \hat{e} = [2.00 \quad 2.00 \quad 1.00]^t$$

which is accurate to the full 3 significant digits!

Exercise 4.1. *Algorithm 4.1 describes iterative improvement. For each step, give the estimate of its computational complexity (its "FLOP count").*

Exercise 4.2. *Show that, when double precision is desired, it can be more efficient for large N to compute the LU factorization in single precision and use iterative refinement instead of using double precision for the factorization and solution. The algorithm can be described as:*

(1) Convert the matrix A to single precision from double precision.
(2) Find the factors L and U in single precision.
(3) Use Algorithm 4.1 to improve the accuracy of the solution. Use A in double precision to compute the double precision residual.

Estimate the FLOP count for the algorithm as outlined above, assuming ten iterations are necessary for convergence. Count each double precision operation as two FLOPs and count each change of precision as one FLOP. Compare this value with the operation count for double precision factorization with a pair of double precision backsolves.

4.2   Vector Norms

> "Intuition is a gift.... . Rare is the expert who combines an informed opinion with a strong respect for his own intuition."
> — G. de Becker, 1997.

Iterative improvement introduces interesting questions like:

- How to measure improvement in an answer?
- How to measure residuals?
- How to quantify ill-conditioning?

The answer to all these questions involves **norms**. A **norm** is a generalization of length and is used to measure the size of a vector or a matrix.

Definition 4.1. Given $x \in \mathbb{R}^N$, a norm of x, $\|x\|$, is a nonnegative real number satisfying

- (Definiteness) $\|x\| \geq 0$ and $\|x\| = 0$ if and only if $x = 0$.
- (Homogeneity) For any real number α and all $x \in \mathbb{R}^N$

$$\|\alpha x\| = |\alpha| \|x\|.$$

- (The triangle inequality) : For all $x, y \in \mathbb{R}^N$

$$\|x + y\| \leq \|x\| + \|y\|.$$

Example 4.3 (Important norms). (i) The Euclidean, or ℓ_2, norm:

$$\|x\|_2 = \sqrt{x \cdot x} = \left(|x_1|^2 + |x_2|^2 + \ldots + |x_N|^2 \right)^{1/2}.$$

(ii) 1-norm or ℓ_1 norm:

$$\|x\|_1 := |x_1| + |x_2| + \ldots + |x_N|.$$

(iii) The max norm or ℓ_∞ norm

$$\|x\|_\infty := \max_{1 \leq j \leq N} |x_j|.$$

(iv) The p-norm or ℓ_p norm: for $1 \leq p < \infty$,

$$\|x\|_p = \left(|x_1|^p + |x_2|^p + \ldots + |x_N|^p \right)^{1/p}.$$

The max norm is called the ℓ_∞ norm because

$$\|x\|_p \to \max_j |x_j|, \quad as \quad p \to \infty.$$

Proposition 4.1 (Norm Equivalence). *For all $x \in \mathbb{R}^N$ have:*

$$\|x\|_\infty \leq \|x\|_1 \leq N \|x\|_\infty.$$

If the number of variables N is large, it is common to redefine these norms to make them independent of n by requiring $\|(1, 1, \ldots, 1)\| = 1$. This gives the perfectly acceptable modifications of (i)–(iv) below:

(i) $\|x\|_{\text{RMS}} := \sqrt{\frac{1}{N} \sum_{j=1}^N x_j^2}$, (the "*Root, Mean, Square* " norm).

(ii) $\|x\|_{AVG} := \frac{1}{N} (|x_1| + \ldots + |x_N|)$, the Average size of the entries.

   The weighted norms $\| \cdot \|_1$ (the average), $\| \cdot \|_2$ (the root mean square) and $\| \cdot \|_\infty$ (the maximum) are by far the most important. Only the $\| \cdot \|_2$ or

RMS norm comes from an inner product. Other weights are possible, such as

$$|||x||| := \sqrt{\sum_{j=1}^{N} \omega_j x_j^2}, \text{ where } \omega_j > 0 \text{ and } \sum_{j=1}^{N} \omega_j = 1.$$

Weighted norms are used in cases where different components have different significance, uncertainty, impact on the final answer, *etc.*

Exercise 4.3. *Show that* $\|x\|_p \to \max_j |x_j|$, *as* $p \to \infty$.

4.2.1   *Norms that come from inner products*

The Euclidean or l_2 norm comes from the usual dot product by

$$\|x\|_2^2 = x \cdot x = \sum |x_i|^2.$$

Dot products open geometry as a tool for analysis and for understanding since the angle[1] between two vectors x, y can be defined through the dot product by

$$\cos(\theta) = \frac{x \cdot y}{\|x\|_2 \|y\|_2}.$$

Thus norms that are induced by dot products are special because they increase the number of tools available for analysis.

Definition 4.2. An inner product on $\mathbb{R}^N$ is a map: $x, y \to \langle x, y \rangle_*$, mapping $\mathbb{R}^N \times \mathbb{R}^N \to \mathbb{R}$ and satisfying the following:

- (definiteness) $\langle x, x \rangle_* \geq 0$ and $\langle x, x \rangle_* = 0$ if and only if $x = 0$.
- (bilinearity) For any real number α, β and all $x, y, z \in \mathbb{R}^N$

$$\langle \alpha x + \beta y, z \rangle_* = \alpha \langle x, z \rangle_* + \beta \langle y, z \rangle_*.$$

- (symmetry) : For all $x, y \in \mathbb{R}^N$

$$\langle x, y \rangle_* = \langle y, x \rangle_*.$$

Proposition 4.2 (Inner product induces a norm). *If* $\langle \cdot, \cdot \rangle_*$ *is an inner product then* $\|x\|_* = \sqrt{\langle x, x \rangle_*}$ *is the norm induced by the inner product.*

---

[1] In statistics this is called the *correlation* between x and y. If the value is 1 the vectors point the same way and are thus perfectly correlated. If its -1 they are said to be anti-correlated.

Since an inner product is a generalization of the usual euclidean dot product it is therefore no surprise that norms and angles can be defined through any given dot product by

$$\text{Induced norm:} \quad ||x||_* = \sqrt{\langle x, x \rangle_*}$$
$$\text{Induced angle:} \quad \cos_*(\theta) = \frac{\langle x, y \rangle_*}{||x||_* \, ||y||_*}.$$

The following definition shows that *orthogonality* has the expected meaning.

Definition 4.3. Vectors x, y are **orthogonal** in the inner product $\langle x, y \rangle_*$ if $\langle x, y \rangle_* = 0$ (and thus the induced angle between them is $\pi/2$). Vectors x, y are **orthonormal** if they are orthogonal and have induced norm one $||x||_* = ||y||_* = 1$. A set of vectors is orthogonal (respectively orthonormal) if elements are pairwise orthogonal (respectively orthonormal).

We have used the subscript $*$ as a place holder in our definition of inner product because it will be convenient to reserve $\langle x, y \rangle$ for the usual euclidean dot product:

$$\langle x, y \rangle := \sum_{j=1}^{N} x_j y_j = x \cdot y.$$

Vectors are operated upon by matrices so the question of how angles can change thereby can be important. There is one special case with an easy answer.

Theorem 4.2. *Let $\langle x, y \rangle$ denote the usual euclidean inner product. If A is an $N \times N$ real, symmetric matrix (i.e., if $A = A^t$ or $a_{ij} = a_{ji}$) then for all x, y*

$$\langle Ax, y \rangle = \langle x, Ay \rangle.$$

More generally, for all x, y and any $N \times N$ matrix A

$$\langle Ax, y \rangle = \langle x, A^t y \rangle.$$

Proof. We calculate (switching the double sum[2])

$$\langle Ax, y \rangle = \sum_{i=1}^{N} \left(\sum_{j=1}^{N} a_{ij} x_j \right) y_i$$

$$= \sum_{j=1}^{N} \left(\sum_{i=1}^{N} a_{ij} y_i \right) x_j = \langle x, A^t y \rangle.$$

$\square$

---

[2]Every proof involving a double sum seems to be done by switching their order then noticing what you get.

The property that $\langle Ax, y \rangle = \langle x, Ay \rangle$ is called *self-adjointness with respect to the given inner product* $\langle \cdot, \cdot \rangle$. If the inner product changes, the matrices that are self-adjoint change and must be redetermined from scratch.

Often the problem under consideration will induce the norm one is forced to work with. One common example occurs with SPD matrices.

Definition 4.4. An $N \times N$ matrix A is symmetric positive definite, SPD for short, if

- A is symmetric: $A^t = A$, and
- A is positive definite: for all $x \neq 0, x^t Ax > 0$.

SPD matrices can be used to induce inner products and norms as follows.

Definition 4.5. Suppose A is SPD. The A inner product and A norm are

$$\langle x, y \rangle_A := x^t Ay, \text{ and } ||x||_A := \sqrt{\langle x, x \rangle_A}.$$

The A inner product is of special importance for solutions of $Ax = b$ when A is SPD. Indeed, using the equation $Ax = b$, $\langle x, y \rangle_A$ can be calculated when A is SPD without knowing the vector x as follows:

$$\langle x, y \rangle_A = x^t Ay = (Ax)^t y = b^t y.$$

Exercise 4.4. *Prove that if $\langle \cdot, \cdot \rangle_*$ is an inner product then $||x||_* = \sqrt{\langle x, x \rangle_*}$ is a norm.*

Exercise 4.5. *Prove that if A is SPD then $\langle x, y \rangle_A := x^t Ay$ is an inner product. Show that A, A^2, A^3, $\cdots$ are self adjoint with respect to the A inner product: $\langle A^k x, y \rangle_A = \langle x, A^k y \rangle_A$.*

Exercise 4.6. *If $\langle \cdot, \cdot \rangle_*$ satisfies two but not all three conditions of an inner product find which conditions in the definition of a norm are satisfied and which are violated. Apply your analysis to $\langle x, y \rangle_A := x^t Ay$ when A is not SPD.*

Exercise 4.7. *The unit ball is $\{x : ||x||_* \leq 1\}$. Sketch the unit ball in $\mathbb{R}^2$ for the 1, 2 and infinity norms. Note that the only ball that looks ball-like is the one for the 2-norm. Sketch the unit ball in the weighted 2 norm induced by the inner product $\langle x, y \rangle := (1/4)x_1 y_1 + (1/9)x_2 y_2$.*

Exercise 4.8. *An $N \times N$ matrix is orthogonal if its columns are N orthonormal (with respect to the usual euclidean inner product) vectors. Show that if O is an orthogonal matrix then $O^T O = I$, and that $||Ox||_2 = ||x||_2$.*

4.3 Matrix Norms

"Wir müssen wissen.

Wir werden wissen."

— David Hilbert (1862–1943) [Engraved on his tombstone in Göttingen.]

It is easy to define a norm on matrices by thinking of an $N \times N$ matrix as just an ordered collection of N^2 real numbers. For example, $\max_{i,j} |a_{ij}|$ is a norm as is the so called Frobenius norm,

$$\|A\|_{\text{Frobenius}} := \sqrt{\sum_{j=1}^{n} \sum_{i=1}^{n} a_{ij}^2} \ .$$

However, most such norms are **not useful. Matrices multiply vectors.** Thus, a useful norm is one which can be used to bound how much a vector grows when multiplied by A. Thus, under $y = Ax$ we seek a notion of $\|A\|$ under which

$$\|y\| = \|Ax\| \leq \|A\|\|x\|.$$

Starting with the essential function a matrix norm must serve and working backwards gives the following definition.

Definition 4.6 (Matrix Norm). Given an $N \times N$ matrix A and a vector norm $\| \cdot \|$, the **induced matrix norm of** A is defined by

$$\|A\| = \max_{x \in \mathbb{R}^N, x \neq 0} \frac{\|Ax\|}{\|x\|}.$$

By this definition, $\|A\|$ is the smallest number such that

$$\|Ax\| \leq \|A\|\|x\| \quad \text{for all} \ \ x \in \mathbb{R}^N.$$

The property that $\|Ax\| \leq \|A\|\|x\|$ for all $x \in \mathbb{R}^N$ is the key to using matrix norms. It also follows easily from the definition of matrix norms that

$$\|I\| = 1.$$

Many features of the induced matrix norm follow immediately from properties of the starting vector norm, such as the following.

Theorem 4.3 (A norm on matrices). *The induced matrix norm is a norm on matrices because if A, B are $N \times N$ matrices and α a scalar, then*

(1) $\|\alpha A\| = |\alpha| \|A\|$.
(2) $\|A\| \geq 0$ *and* $\|A\| = 0$ *if and only if* $A \equiv 0$.
(3) $\|A + B\| \leq \|A\| + \|B\|$.

Proof. Exercise!   □

Other features follow from the fact that matrix norms split products apart, such as the following.

Theorem 4.4 (Properties of Matrix Norms). *Let* A, B *be* $N \times N$ *matrices and* α *a scalar. Then*

(1) $\|Ax\| \leq \|A\| \|x\|$.
(2) $\|AB\| \leq \|A\| \|B\|$.
(3) *If* A *is invertible, then for all* x

$$\frac{\|x\|}{\|A^{-1}\|} \leq \|Ax\| \leq \|A\| \|x\|.$$

(4) $\|A\| = \max_{\|x\|=1, x \in \mathbb{R}^N} \|Ax\|$.
(5) $\|A^{-1}\| \geq \frac{1}{\|A\|}$.
(6) *For any* $N \times N$ *matrix* A *and* $\|\cdot\|$ *any matrix norm:*

$$|\lambda(A)| \leq \|A\|.$$

Proof. We will prove some of these to show how $\|Ax\| \leq \|A\| \|x\|$ is used in getting bounds on the action if a matrix. For example, note that $A^{-1}Ax = x$. Thus

$$\|x\| \leq \|A^{-1}\| \|Ax\|,$$

so

$$\frac{\|x\|}{\|A^{-1}\|} \leq \|Ax\| \leq \|A\| \|x\|.$$

For (5), $A^{-1}A = I$ so $\|I\| = 1 \leq \|A^{-1}\| \|A\|$ (using (2)), and $\|A^{-1}\| \geq 1/\|A\|$. For number 6, since $A\phi = \lambda\phi$. Thus $|\lambda| \|\phi\| = \|A\phi\| \leq \|A\| \|\phi\|$. □

Remark 4.1. The fundamental property that $\|AB\| \leq \|A\| \|B\|$ for all A, B shows the key to using it to structure proofs. As a first example, consider the above proof of $\frac{\|x\|}{\|A^{-1}\|} \leq \|Ax\|$. How is one to arrive at this proof? To begin rearrange so it becomes $\|x\| \leq \|A^{-1}\| \|Ax\|$. The top (upper) side of such an inequality must come from splitting a product apart. This suggests starting with $\|A^{-1}Ax\| \leq \|A^{-1}\| \|Ax\|$. Next observe the LHS is just $\|x\|$.

The matrix norms is defined in a nonconstructive way. However, there are a few special cases when the norm can be calculated:

- $\|A\|_\infty$ is calculable; it is the **maximum row sum**. It has value

$$\|A\|_\infty = \max_{1 \le i \le n} \sum_{j=1}^{n} |a_{ij}|.$$

- $\|A\|_1$ is calculable; it is the **maximum column sum**. It has value

$$\|A\|_1 = \max_{1 \le j \le n} \sum_{i=1}^{n} |a_{ij}|.$$

- If $A = A^t$ is symmetric then $\|A\|_2$ is calculable. $\|A\|_2 = \max\{|\lambda| : \lambda$ is an eigenvalue of $A\}$
- For general $A_{N \times N}$, $\|A\|_2$ is calculable. It is the largest singular value of A, or the square root of the largest eigenvalue of $A^t A$:

$$\|A\|_2 = \sqrt{\max\{\lambda : \lambda \text{ is an eigenvalue of } A^t A\}}.$$

Example 4.4. Let A be the 2×2 matrix

$$A = \begin{bmatrix} +1 & -2 \\ -3 & +4 \end{bmatrix}.$$

We calculate

$$\|A\|_\infty = \max\{|1| + |-2|, |-3| + |4|\} = \max\{3, 7\} = 7,$$
$$\|A\|_1 = \max\{|1| + |-3|, |-2| + |4|\} = \max\{4, 6\} = 6.$$

Since A is not symmetric, we can calculate the 2 norm either from the singular values of A or directly from the definition of A. For the 2×2 case we can do the latter. Recall

$$\|A\|_2 = \max\{\|Ax\|_2 : \|x\|_2 = 1\}.$$

Every unit vector in the plane can be written as

$$x = (\cos\theta, \sin\theta)^t \text{ for some } \theta.$$

We compute

$$Ax = (\cos\theta - 2\sin\theta, -3\cos\theta + 4\sin\theta)^t,$$
$$\|Ax\|_2^2 = (\cos\theta - 2\sin\theta)^2 + (-3\cos\theta + 4\sin\theta)^2 \equiv f(\theta).$$

Thus

$$\|Ax\|_2 = \sqrt{\max_{0 \le \theta \le 2\pi} f(\theta)},$$

which is a calculus problem (Exercise 4.9).

Alternately we can compute

$$A^t A = \begin{bmatrix} 10 & -14 \\ -14 & 20 \end{bmatrix}.$$

Then we calculate the eigenvalues of $A^t A$ by

$$0 = \det \begin{bmatrix} 10 - \lambda & -14 \\ -14 & 20 - \lambda \end{bmatrix} \Rightarrow (10 - \lambda)(20 - \lambda) - 14^2 = 0,$$

whereupon

$$||Ax||_2 = \sqrt{\max\{\lambda_1, \lambda_2\}}.$$

Exercise 4.9. *Complete the above two calculations of* $||A||_2$.

Exercise 4.10. *Calculate the* $1, 2$ *and* ∞ *norms of* A

$$A = \begin{bmatrix} 1 & -3 \\ -4 & 7 \end{bmatrix}$$

and

$$A = \begin{bmatrix} 1 & -3 \\ -3 & 7 \end{bmatrix}.$$

Exercise 4.11. *Show that an orthogonal change of variables preserves the* 2 *norm:* $||Ox||_2 = ||x||_2$ *if* $O^t O = I$.

Exercise 4.12. *Prove that, for any induced matrix norm,*

$$||I|| = 1, \text{ and}$$
$$||A^{-1}|| \geq 1/||A||.$$

4.3.1 A few proofs

We next prove these claimed formulas.

Theorem 4.5 (Calculation of 2 norm of a symmetric matrix). *If* $A = A^t$ *is symmetric then* $||A||_2$ *is given by*

$$||A||_2 = \max\{|\lambda| : \lambda \text{ is an eigenvalue of } A\}.$$

Proof. If A is symmetric then it is diagonalizable by a real orthogonal matrix[3] O:

$$A = O^t \Lambda O, \text{ where } O^t O = I \text{ and } \Lambda = diag(\lambda_i).$$

---

[3] Recall that an orthogonal matrix is one where the columns are mutually orthonormal. This implies $O^t O = I$.

We then have by direct calculation

$$\|A\|_2^2 = \max_{x \in \mathbb{R}^N, x \neq 0} \left(\frac{\|Ax\|_2}{\|x\|_2} \right)^2$$

$$= \max_{x \in \mathbb{R}^N, x \neq 0} \frac{\langle Ax, Ax \rangle}{\langle x, x \rangle} = \max_{x \in \mathbb{R}^N, x \neq 0} \frac{\langle O^t \Lambda O x, O^t A O x \rangle}{\langle x, x \rangle}$$

$$= \max_{x \in \mathbb{R}^N, x \neq 0} \frac{\langle \Lambda O x, \Lambda O x \rangle}{\langle x, x \rangle}.$$

Now change variables on the RHS by $y = Ox, x = O^t y$. An elementary calculation (Exercise 4.11) shows that an orthogonal change of variables preserves the 2 norm: $\|Ox\|_2 = \|x\|_2$. This gives

$$\|A\|_2^2 = \max_{y \in \mathbb{R}^N, y = Ox \neq 0} \frac{\langle \Lambda y, \Lambda y \rangle}{\|x\|_2^2} = \max_{y \in \mathbb{R}^N, y \neq 0} \frac{\langle \Lambda y, \Lambda y \rangle}{\langle y, y \rangle}$$

$$= \max_{y \in \mathbb{R}^N, y \neq 0} \frac{\sum_i \lambda_i^2 y_i^2}{\sum_i y_i^2} = (|\lambda|_{\max})^2.$$

$\square$

The proof the formulas for the 1 and infinity norms are a calculation.

Theorem 4.6 (Matrix 1-norm and ∞-norm). *We have*

$$\|A\|_\infty = \max_{1 \leq i \leq N} \sum_{j=1}^{N} |a_{ij}|,$$

$$\|A\|_1 = \max_{1 \leq j \leq N} \sum_{i=1}^{N} |a_{ij}|.$$

Proof. Consider $\|A\|_1$. Partition A by column vectors (so $\vec{a_j}$ denotes the jth column) as

$$A = [\vec{a_1} | \vec{a_2} | \vec{a_3} | \cdots | \vec{a_N}].$$

Then we have

$$Ax = x_1 \vec{a_1} + \cdots + x_N \vec{a_N}.$$

Thus, by the triangle inequality

$$\|Ax\|_1 \leq |x_1| \cdot \|\vec{a_1}\|_1 + \cdots + |x_N| \cdot \|\vec{a_N}\|_1$$

$$\leq \left(\sum_i |x_i| \right) \left(\max_j \|\vec{a_j}\|_1 \right) = \|x\|_1 \left(\max_{1 \leq j \leq N} \sum_{i=1}^{N} |a_{ij}| \right).$$

Dividing by $\|x\|_1$ we have thus

$$\|A\|_1 \le \max_{1\le j\le N} \sum_{i=1}^{N} |a_{ij}|.$$

To prove equality, we take j^* to be the index (of the largest column vector) for which $\max_j \|\overrightarrow{a_j}\|_1 = \|\overrightarrow{a_{j^*}}\|_1$ and choose $x = e_{j^*}$. Then

$$Ae_{j^*} = 1\overrightarrow{a}_{j^*} \quad \text{and}$$

$$\frac{\|Ae_{j^*}\|_1}{\|e_{j^*}\|_1} = \frac{\|a_{j^*}\|_1}{\|e_{j^*}\|_1} = \|a_{j^*}\|_1 = \max_{1\le j\le N} \sum_{i=1}^{N} |a_{ij}|.$$

We leave the proof for $\|A\|_\infty = \max_{1\le i\le N} \sum_{j=1}^{N} |a_{ij}|$ as an exercise. $\square$

"He (Gill) climbed up and down the lower half of the rock over and over, memorizing the moves... He says that '... going up and down, up and down eventually... your mind goes blank and you climb by well cultivated instinct'. "
— J. Krakauer, from his book *Eiger Dreams*.

Exercise 4.13. *Show that $\|A\|_\infty = \max_{1\le i\le n} \sum_{j=1}^{n} |a_{ij}|$. Hint:*

$$(Ax)_i = |\sum_{j=1}^{n} a_{ij}x_j| \le \sum_{j=1}^{n} |a_{ij}||x_j| \le$$

$$\left(\max_j |x_j|\right) \sum_{j=1}^{n} |a_{ij}| = \|x\|_\infty \cdot (Sum\ of\ row\ i).$$

Exercise 4.14. *Show that for A an $N \times N$ symmetric matrix*

$$\|A\|_{Frobenius}^2 = trace(A^tA) = \sum_i \lambda_i^2.$$

Exercise 4.15. *If $\|\cdot\|$ is a vector norm and U an $N \times N$ nonsingular matrix, show that $\|x\|_* := \|Ux\|$ is a vector norm. When $\|\cdot\| = \|\cdot\|_2$, find a formula for the matrix norm induced by $\|\cdot\|_*$.*

4.4 Error, Residual and Condition Number

We [he and Halmos] share a philosophy about linear algebra: we think basis-free, we write basis-free, but when the chips are down we close the office door and compute with matrices like fury.

— Kaplansky, Irving, Paul Halmos: Celebrating 50 Years of Mathematics.

"What is now proved was once only imagin'd."

— W. Blake, *The Marriage of Heaven and Hell*, 1790-3.

If we solve $Ax = b$ and produce an approximate solution $\widehat{x}$, then the fundamental equation of numerical linear algebra, $Ae = r$, links error and residual, where

$$error := e = x - \widehat{x},$$
$$residual := r = b - A\widehat{x}.$$

Recall that, while $e = 0$ if and only if $r = 0$, there are cases where small residuals and large errors coexist. For example, the point $P = (0.5, 0.7)$ and the 2×2 linear system

$$x - y = 0$$
$$-0.8x + y = 1/2$$

are plotted in Figure 4.1.

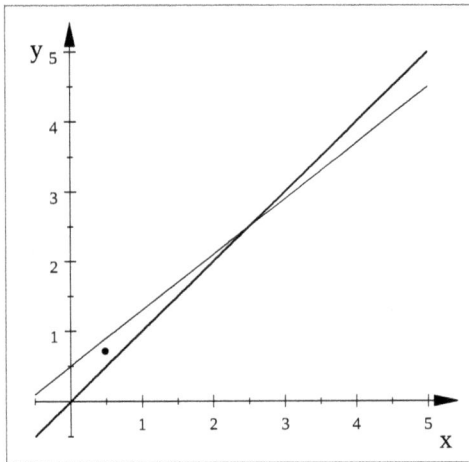

Fig. 4.1   Lines L1, L2 and the point P.

The point P is close to both lines so the residual of P is small. However, it is far from the solution (the lines' intersections). We have seen the following *qualitative* features of this linkage:

- If A is invertible then $r = 0$ if and only if $e = 0$.
- The residual is computable but e is not exactly computable in a useful sense since solving $Ae = r$ for e is comparable to solving $Ax = b$ for x.
- If A is well conditioned then $||r||$ and $||e||$ are comparable in size.
- If A is ill conditioned then $||r||$ can be small while $||e||$ is large.
- $det(A)$ cannot be the right way to quantity this connection. Starting with $Ae = r$ we have $(\alpha A)\,e = \alpha r$ and rescaling can make $det(\alpha A) = 1$ using $det(\alpha A) = \alpha^n det(A)$.

This section makes a precise quantitative connection between the size of errors and residuals and, in the process, quantify conditioning.

Definition 4.7. Let $|| \cdot ||$ be a matrix norm. Then the condition number of the matrix A induced by the vector norm $|| \cdot ||$ is

$$cond_{||\cdot||}(A) := ||A^{-1}||\,||A||.$$

Usually the norm in question will be clear from the context in which cond(A) occurs so usually the subscript of the norm is omitted. The condition number of a matrix is also often denoted by the Greek letter kappa:

$$\kappa(A) = cond(A) = \text{ condition number of } A.$$

Theorem 4.7 (Relative Error $\leq cond(A)\times$Relative Residual). *Let $Ax = b$ and let $\hat{x}$ be an approximation to the solution x. With $r = b - A\hat{x}$*

$$\frac{||x - \hat{x}||}{||x||} \leq \text{cond}(A)\frac{||r||}{||b||}. \qquad (4.2)$$

Proof. Begin with $Ae = r$ then $e = A^{-1}r$. Thus,

$$||e|| = ||A^{-1}r|| \leq ||A^{-1}||\,||r||. \qquad (4.3)$$

Since $Ax = b$ we also know $||b|| = ||Ax|| \leq ||A||\,||x||$. Dividing the smaller side of (4.3) by the larger quantity and the larger side of (4.3) by the smaller gives

$$\frac{||e||}{||A||\,||x||} \leq \frac{||A^{-1}||\,||r||}{||b||}.$$

Rearrangement proves the theorem.   □

Equation (4.2) quantifies ill-conditioning: the larger cond(A) is, the more ill-conditioned the matrix A.

Remark 4.2. The manipulations in the above proof are typical of ones in numerical linear algebra and the actual result is a cornerstone of the field.

Note the pattern: we desire an inequality $||error|| \leq ||terms||$. Thus, we begin with an equation $error = product$ then take norms of both sides.

Example 4.5. $||I|| = 1$ so $\text{cond}(I) = 1$ in any *induced* matrix norm. Similarly for an orthogonal matrix $\text{cond}_2(O) = 1$.

Example 4.6. Let

$$A = \begin{bmatrix} 1.01 & 0.99 \\ 0.99 & 1.01 \end{bmatrix}$$

then $||A||_\infty = 2$. Since for any 2×2 matrix[4]

$$\begin{bmatrix} a & b \\ c & d \end{bmatrix}^{-1} = \frac{1}{\det(A)} \begin{bmatrix} d & -b \\ -c & a \end{bmatrix},$$

we can calculate A^{-1} exactly

$$A^{-1} = \begin{bmatrix} 25.25 & -24.75 \\ -24.75 & 25.25 \end{bmatrix}.$$

Hence, $||A^{-1}||_\infty = 50$. Thus $\text{cond}(A) = 100$ and errors can be (at most) $100\times$ larger than residuals.

Example 4.7. Let A be as above and $b = (2 \ 2)^t$

$$A = \begin{bmatrix} 1.01 & 0.99 \\ 0.99 & 1.01 \end{bmatrix}, \quad b = \begin{bmatrix} 2 \\ 2 \end{bmatrix},$$

so $x = (1 \ 1)^t$. Consider $\hat{x} = (2 \ 0)^t$. The error is $e = x - \hat{x} = (-1 \ 1)^t$ and $||e||_\infty = 1$. The residual of $\hat{x}$ is

$$r = b - A\hat{x} = \begin{bmatrix} -0.02 \\ -0.02 \end{bmatrix}.$$

As $||r||_\infty = 0.02$ we see

$$\frac{||r||}{||b||} = \frac{0.02}{2} \quad and \quad \frac{||e||}{||x||} = \frac{1}{1}$$

so the error is exactly $100\times$ larger than the residual!

Example 4.8 (The Hilbert Matrix). The $N \times N$ Hilbert matrix $H_{N \times N}$ is the matrix with entries

$$H_{ij} = \frac{1}{i + j - 1}, \quad 1 \leq i, j \leq n.$$

This matrix is extremely ill conditioned even for quite moderate values on n.

---

[4]This formula for the inverse of a 2×2 matrix is handy for constructing explicit 2×2 examples with various features. It does not hold for 3×3 matrices.

Example 4.9. Let $x = (1.0, 1.0)^t$ and

$$A = \begin{bmatrix} 1.000 & 2.000 \\ .499 & 1.001 \end{bmatrix}, \quad b = \begin{bmatrix} 3.00 \\ 1.50 \end{bmatrix}.$$

Given $\hat{x} = (2.00 \ 0.500)^t$ we calculate $r = b - A\hat{x} = (0, 0.0015)^t$ which is "small" in both an absolute and relative sense. As in the previous example we can find A^{-1} and then calculate $\|A^{-1}\|_\infty$. We find

$$\|A\|_\infty = 3, \quad \|A^{-1}\|_\infty = \left\| \begin{bmatrix} 333.67 & -666.67 \\ -166.67 & 333.33 \end{bmatrix} \right\|_\infty = 1000.34.$$

Thus,

$$\text{cond}(A) = 3001.02$$

and the relative residual can be 3000× smaller that the relative error. Indeed, we find

$$\frac{\|x - \hat{x}\|_\infty}{\|x\|_\infty} = 1, \text{ and } \frac{\|r\|_\infty}{\|b\|_\infty} = 0.0045.$$

Example 4.10. Calculate $\text{cond}(H_N)$, for $N = 2, 3, \cdots, 13$. (This is easily done in MATLAB.) Plot $\text{cond}(H)$ vs N various ways and try to find its growth rate. Do a literature search and find it.

Exercise 4.16. *Suppose $Ax = b$ and that $\hat{x}$ is an approximation to x. Prove the result of Theorem 4.7 that*

$$\frac{\|x - \hat{x}\|}{\|x\|} \le \text{cond}_{\|\cdot\|}(A) \frac{\|r\|}{\|b\|}.$$

Prove the associated lower bound

$$\frac{\|x - \hat{x}\|}{\|x\|} \ge [1/\text{cond}_{\|\cdot\|}(A)] \frac{\|r\|}{\|b\|}.$$

Hint: Think about which way the inequalities must go to have the error on top.

Exercise 4.17. *Let A be the following 2×2 symmetric matrix. Find the eigenvalues of A and the 2 norm of A and $\text{cond}_2(A)$:*

$$A = \begin{bmatrix} 1 & 2 \\ 2 & 4 \end{bmatrix}.$$

Exercise 4.18. *Show that for any square matrix (not necessarily symmetric) $\text{cond}_2(A) \ge |\lambda|_{\max}/|\lambda|_{\min}$.*

Exercise 4.19.

(1) If $cond(A) = 10^6$ and you solve $Ax = b$ on a computer with 7 significant digits (base 10), **What** is the expected number of significant digits of accuracy of the solution?

(2) Let $Ax = b$ and $\widehat{x}$ let be some approximation to x, $e = x - \widehat{x}$, $r = b - A\widehat{x}$.

(3) **Show that** $Ae = r$ and **explain at least 3** places where this equation is useful or important in numerical linear algebra.

(4) **Show that**:

$$||e||/||x|| \leq cond(A)||r||/||b||.$$

4.5 Backward Error Analysis

One of the principal objects of theoretical research in my department of knowledge is to find the point of view from which the subject appears in its greatest simplicity.
— Gibbs, Josiah Willard (1839–1903)

For many problems in numerical linear algebra results of the following type have been proven by meticulously tracking through computer arithmetic step by step in the algorithm under consideration. It has been verified in so many different settings that it has become something between a meta theorem and a philosophical principle of numerical linear algebra.

Basic Result of Backward Error Analysis.[5] **The result** $\hat{x}$ **of solving** $Ax = b$ **by Gaussian elimination in finite precision arithmetic subject to rounding errors is precisely the same as the exact solution of a perturbed problem**

$$(A + E)\hat{x} = b + f \tag{4.4}$$

where

$$\frac{||\mathbf{E}||}{||\mathbf{A}||}, \frac{||f||}{||b||} = O(\text{machine precision}).$$

First we consider the effect of perturbations to the matrix A. Since the entries in the matrix A are stored in finite precision, these errors occur even if all subsequent calculations in Gaussian elimination were done in infinite precision arithmetic.

---

[5]This has been proven for most matrices A. There are a few rare types of matrices for which it is not yet proven and it is an open question if it holds for all matrices (i.e., for those rare examples) without some adjustments.

Theorem 4.8 (Effect of storage errors in A). *Let A be an $N \times N$ matrix. Suppose $Ax = b$ and $(A + E)\hat{x} = b$. Then,*

$$\frac{\|x - \hat{x}\|}{\|\hat{x}\|} \leq cond(A)\frac{\|E\|}{\|A\|}.$$

Proof. The proof has a well defined strategy that we shall use in other proofs:

Step 1: By subtraction get an equation for the error driven by the perturbation:

$$Ax = b$$
$$-(A\hat{x} = b - E\hat{x})$$
$$\_\_\_\text{subtract}\_\_\_\_$$
$$A(x - \hat{x}) = E\hat{x}$$
$$x - \hat{x} = A^{-1}E\hat{x}.$$

Step 2: Bound error by RHS:

$$\|x - \hat{x}\| = \|A^{-1}E\hat{x}\| \leq \|A^{-1}\| \cdot \|E\| \cdot \|\hat{x}\|.$$

Step 3: Rearrange to write in terms of relative quantities and condition numbers:

$$\frac{\|x - \hat{x}\|}{\|\hat{x}\|} \leq cond(A)\frac{\|E\|}{\|A\|}.$$

$\square$

Theorem 4.9 (Effect of storage errors in b). *Let $Ax = b$ and $A\hat{x} = b + f$. Then*

$$\frac{\|x - \hat{x}\|}{\|x\|} \leq cond(A)\frac{\|f\|}{\|b\|}.$$

Proof. Since $A\hat{x} = Ax + f$, $x - \hat{x} = -A^{-1}f$, $\|x - \hat{x}\| \leq \|A^{-1}\|\|f\| = cond(A)\frac{\|f\|}{\|A\|} \leq cond(A)\frac{\|f\|}{\|b\|}\|x\|$, because $\|b\| \leq \|A\|\|x\|$. $\square$

Remark 4.3 (Interpretation of $cond(A)$). When E is due to roundoff errors $\|E\|/\|A\| = O$ (machine precision). Then these results say that: $cond(A)$ *tells you how many significant digits are lost (worst case) when solving $Ax = b$.* As an example, if machine precision carries 7 significant digits, $\|E\|/\|A\| = O(10^{-7})$, and if $cond(A) = 10^5$ then $\hat{x}$ will have at least $7 - 5 = 2$ significant digits.

Other properties of $cond(A)$:

- $cond(A) \geq 1$ and $cond(I) = 1$.
- Scaling A does not influence $cond(A)$:

$$cond(\alpha A) = cond(A), \quad \text{for any } \alpha \neq 0.$$

- $cond(A)$ depends on the norm chosen but usually it is of the same order of magnitude for different norms.
- $cond(A)$ is **not** related to $\det(A)$. For example, scaling changes $\det(A)$ but not $cond(A)$:

$$\det(\alpha A) = \alpha^n \det(A) \quad \text{but} \quad cond(\alpha A) = cond(A).$$

- If A is symmetric then

$$cond_2(A) = |\lambda|_{max}/|\lambda|_{min}.$$

- If A is symmetric, positive definite and $\| \cdot \| = \| \cdot \|_2$, then $cond(A)$ equals the **spectral condition number**, $\lambda_{max}/\lambda_{min}$

$$cond_2(A) = \lambda_{max}/\lambda_{min}.$$

- $cond(A) = cond(A^{-1})$.
- We shall see in a later chapter that the error in eigenvalue and eigenvector calculations is also governed by $cond(A)$.

The most important other result involving $cond(A)$ is for the perturbed system when there are perturbations in both A and b.

4.5.1 *The general case*

Say what you know, do what you must, come what may.
— Sonja Kovalevsky, [Motto on her paper "On the Problem of the Rotation of a Solid Body about a Fixed Point"].

We show next that the error in

$$(A + E)\hat{x} = b + f \text{ compared to the true system:} \quad Ax = b$$

is also governed by $cond(A)$. This requires some technical preparation.

Lemma 4.1 (Spectral localization). *For any $N \times N$ matrix B and $\| \cdot \|$ any matrix norm:*

$$|\lambda(B)| \leq \|B\|.$$

Proof. $B\phi = \lambda\phi$. Thus $|\lambda|\|\phi\| = \|B\phi\| \leq \|B\|\|\phi\|$.   □

This result holds for any matrix norm. Thus, various norms of A can be calculated and the smallest used are an inclusion radius for the eigenvalues of A.

Theorem 4.10. *Let $B_{N \times N}$. Then we have*

$$\lim_{n \to \infty} B^n = 0 \text{ if and only if there exists a norm } \|\cdot\| \text{ with } \|B\| < 1.$$

Proof. We prove $\|B\| < 1 \Rightarrow \|B^n\| \to 0$. This is easy. The other direction will be proven later.[6] We have that $\|B^2\| = \|B \cdot B\| \leq \|B\| \cdot \|B\| = \|B\|^2$. By induction it follows that $\|B^n\| \leq \|B\|^n$ and thus

$$\|B^n\| \leq \|B\|^n \to 0 \text{ as } n \to \infty.$$

□

We shall use the following special case of the spectral mapping theorem.

Lemma 4.2. *The eigenvalues of $(I - B)^{-1}$ are $(1 - \lambda)^{-1}$ where λ is an eigenvalue of B.*

Proof. Let $B\phi = \lambda\phi$. Then, $\phi - B\phi = \phi - \lambda\phi$ and $(I - B)\phi = (1 - \lambda)\phi$. Inverting we see that $(1 - \lambda)^{-1}$ is an eigenvalue of $(I - B)^{-1}$. Working backwards (with details left as an exercise) it follows similarly that $(1-\lambda)^{-1}$ an eigenvalue of $(I - B)^{-1}$ implies λ is an eigenvalue of B.   □

Theorem 4.11 (The Neumann Lemma). *Let $B_{N \times N}$ be given, with $\|B\| < 1$. Then $(I - B)^{-1}$ exists and*

$$(I - B)^{-1} = \lim_{N \to \infty} \left(\sum_{\ell=0}^{N} B^\ell \right).$$

Proof. IDEA OF PROOF: Just like summing a geometric series:

$$S = 1 + \alpha + \alpha^2 + \cdots + \alpha^N$$
$$\alpha S = \alpha + \cdots + \alpha^N + \alpha^{N+1}$$
$$(1 - \alpha)S = 1 - \alpha^{N+1}.$$

---

[6]Briefly: Exercise 4.22 shows that given a matrix B and any $\varepsilon > 0$, there exists a norm within ε of $spr(B)$. With this result, if there does not exist a norm with $\|B\| < 1$, then there is a $\lambda(B)$ with $|\lambda| = spr(B) > 1$. Picking $x = $ eigenvector of λ, we calculate: $|B^n x| = |\lambda^n x| \to \infty$.

To apply this idea, note that since $|\lambda| \leq \|B\|$, $|\lambda| < 1$. Further, $\lambda(I - B) = 1 - \lambda(B)$ by *the spectral mapping theorem.*[7] Since $|\lambda(B)| < 1$, $\lambda(I - B) \neq 0$ and $(I - B)^{-1}$ exists. We verify that the inverse is as claimed. To begin, note that

$$(I - B)(I + B + \cdots + B^N) = I - B^{N+1}.$$

Since $B^N \to 0$ as $N \to \infty$

$$I + B + \cdots + B^N =$$
$$= (I - B)^{-1}(I - B^{N+1}) = (I - B)^{-1} - (I - B)^{-1}BB^N \to (I - B)^{-1}.$$

$\square$

As an application of the Neumann lemma we have the following.

Corollary 4.1 (Perturbation Lemma). *Suppose A is invertible and $\|A^{-1}\|\|E\| < 1$, then $A + E$ is invertible and*

$$\|(A + E)^{-1}\| \leq \frac{\|A^{-1}\|}{1 - \|A^{-1}\|\|E\|}.$$

Exercise 4.20. *Prove this corollary.*

The ingredients are now in place. We give the proof of the general case.

Theorem 4.12 (The General Case). *Let*

$$Ax = b, \quad (A + E)\hat{x} = b + f.$$

Assume A^{-1} exists and

$$\|A^{-1}\|\|E\| = cond(A)\frac{\|E\|}{\|A\|} < 1.$$

Then

$$\frac{\|x - \hat{x}\|}{\|x\|} \leq \frac{cond(A)}{1 - \|A^{-1}\|\|E\|} \left\{ \frac{\|E\|}{\|A\|} + \frac{\|f\|}{\|b\|} \right\}.$$

---

[7] An elementary proof is because the eigenvalues of $\lambda(B)$ are roots of the polynomial $\det(\lambda I - B) = -\det((1 - \lambda)I - (I - B))$.

Proof. The proof uses same ideas but is a bit more delicate in the order of steps. First,[8]

$$Ax = b \Longleftrightarrow (A + E)x = b + Ex$$
$$\underline{(A + E)\hat{x} = b + f}$$
$$(A + E)e = Ex - f$$
$$e = (A + E)^{-1}(Ex - f)$$

$$\|e\| \le \|(A + E)^{-1}\|(\|E\|\|x\| + \|f\|).$$

Now

$$Ax = b \text{ so } \begin{cases} x = A^{-1}b, \ \|x\| \le \|A^{-1}\|\|b\| \\ \|b\| \le \|A\|\|x\|, \text{ and } \|x\| \ge \|A\|^{-1}\|b\|. \end{cases}$$

Thus,

$$\frac{\|e\|}{\|x\|} \le \|(A + E)^{-1}\| \left(\|E\|\frac{\|x\|}{\|x\|} + \frac{\|f\|}{\|x\|} \right)$$

$$\frac{\|e\|}{\|x\|} \le \|(A + E)^{-1}\| \left(\|E\| + \underbrace{\|A\|}_{\text{factor out this A}} \frac{\|f\|}{\|b\|} \right)$$

$$\frac{\|e\|}{\|x\|} \le \|A\|\|(A + E)^{-1}\| \left(\frac{\|E\|}{\|A\|} + \frac{\|f\|}{\|b\|} \right).$$

Finally, rearrange terms after using

$$\|(A + E)^{-1}\| \le \frac{\|A^{-1}\|}{1 - \|A^{-1}\|\|E\|}.$$

$\square$

Remark 4.4 (How big is the RHS?). If $\|A^{-1}\|\|E\| \ll 1$, we can estimate (e.g., $\frac{1}{1-\alpha} \simeq 1 + \alpha$)

$$\frac{1}{1 - \|A^{-1}\|\|E\|} \sim 1 + \|A^{-1}\|\|E\| = 1 + small$$

so that up to $O(\|A^{-1}\|\|E\|)$ the first order error is governed by $cond(A)$.

[8]The other natural way to start is to rewrite

$$Ax = b \Longleftrightarrow (A + E)x = b + Ex$$
$$\underline{A\hat{x} = b + f - E\hat{x}}$$
$$e = A^{-1}(f - E\hat{x}).$$

Since there are 2 natural starting points, the strategy is to try one and if it fails, figure out why then try the other.

Remark 4.5 (Non-symmetric matrices). The spectral condition number can be deceptive for non-symmetric matrices. Since $\|A\| \geq |\lambda(A)|$ for each of the eigenvalues $\lambda(A)$ of A, $\|A\| \geq |\lambda|_{\max}(A)$ and $\|A^{-1}\| \geq |\lambda(A^{-1})|_{\max} = 1/|\lambda(A)|_{\min}$. We thus have

$$cond(A) \geq \frac{|\lambda(A)|_{\max}}{|\lambda(A)|_{\min}}$$

i.e., spectral condition number $\leq$ condition number. For example, for A and B below, $cond_2(A) = cond_2(B) = O(10^5)$ but we calculate

$$A = \begin{bmatrix} 1 & -1 \\ 1 & -1.00001 \end{bmatrix}, \text{ and } B = \begin{bmatrix} 1 & -1 \\ -1 & 1.00001 \end{bmatrix},$$

$$\frac{|\lambda|_{\max}(A)}{|\lambda|_{\min}(A)} \sim 1, \text{ while } \frac{|\lambda|_{\max}(B)}{|\lambda|_{\min}(B)} \sim 4 \cdot 10^5.$$

There are many, other results related to Theorem 4.12. For example, all the above upper bounds as relative errors can be complemented by lower bounds, such as the following.

Theorem 4.13. *Let $Ax = b$. Given $\hat{x}$ let $r = b - A\hat{x}$. Then,*

$$\frac{\|x - \hat{x}\|}{\|x\|} \geq \frac{1}{cond(A)} \frac{\|r\|}{\|b\|}.$$

Exercise 4.21. *Prove the theorem.*

The relative distance of a matrix A to the closest non-invertible matrix is also related to $cond(A)$. A proof due to Kahan[9] is presented in Exercise 4.22.

Theorem 4.14 (Distance to nearest singular matrix). *Suppose A^{-1} exists. Then,*

$$\frac{1}{cond(A)} = \min\left\{ \frac{\|A - B\|}{\|A\|} : \det(B) = 0 \right\}.$$

Exercise 4.22. *Theorem 4.14 is a remarkable result. One proof due to Kahan depends on ingenious choices of particular vectors and matrices.*

---

[9]W. Kahan, Numerical linear algebra, Canadian Math. Bulletin, 9 (1966), pp. 757–801. Kahan attributes the theorem to "Gastinel" without reference, but does not seem to be attributing the proof. Possibly the Gastinel reference is: Noël Gastinel, Matrices du second degré et normes générales en analyse numérique linéaire, Publ. Sci. Tech. Ministére de l'Air Notes Tech. No. 110, Paris, 1962.

(1) Show that if B is singular, then

$$\frac{1}{cond(A)} \leq \frac{\|A - B\|}{\|A\|}.$$

Hint: If B is singular, there is an x so that $Bx = 0$ and $(A - B)x = Ax$.
Hence $\|A - B\| \geq \|(A - B)x\|/\|x\|$.
(2) Show that there is a matrix B with

$$\frac{1}{cond(A)} = \frac{\|A - B\|}{\|A\|}.$$

Hint: Show that it is possible to choose a vector y so that $\|A^{-1}y\| = \|A^{-1}\|\|y\| \neq 0$, set $w = (A^{-1}y)/\|A^{-1}y\|^2$ and set $B = A - yw^t$.[10]

Example 4.11 (Perturbations of the right hand side b). Consider
the linear system with exact solution $(1, 1)$:

$$3x_1 + 4x_2 = 7$$
$$5x_1 - 2x_2 = 3$$

and let $f = (0.005, -0, 009)^t$ so the RHS is changed to $b+f = (7.005, 2.991)$.
The solution is now

$$\hat{x} = (0.999 \quad 1.002)^t.$$

Since a small change in the RHS produced a corresponding small change in
the solution we have evidence that

$$\begin{bmatrix} 3 & 4 \\ 5 & -2 \end{bmatrix}$$

is well-conditioned.

Now modify the matrix to get the system (with exact solution still
$(1, 1)^t$)

$$x_1 + x_2 = 2$$
$$1.01x_1 + x_2 = 2.01.$$

This system is **ill-conditioned.** Indeed, change the RHS a little bit to

$$b + f = (2.005, 2.005)^t.$$

The new solution $\hat{x}$ is changed a lot to

$$\hat{x} = (0, 2.005)^t.$$

---

[10]Recall that $w^t y = \langle w, y \rangle$ is a scalar but that yw^t, the "outer product," is a matrix.

Example 4.12 (Changes in the coefficients). Suppose the coefficients of the system (solution $= (1,1)^t$)

$$1x_1 + 1x_2 = 2$$
$$1.01x_1 + 1x_2 = 2.01$$

are changed slightly to read

$$1x_1 + 1x_2 = 2$$
$$1.0001x_1 + 1x_2 = 2.001.$$

Then, the exact solution changes wildly to

$$\hat{x} = (100, -98)^t.$$

We still have a very small residual in the perturbed system

$$r_1 = 2 - (1 \cdot 100 + 1 \cdot (-98)) = 0$$
$$r_2 = 2.001 - (1.0001 \cdot 100 + 1 \cdot (-98)) = -0.009.$$

Example 4.13 (cond(A) and det(A) not related). Let ε denote a small positive number. The matrix A below is ill conditioned and its determinant is ε thus near zero:

$$\begin{bmatrix} 1 & 1 \\ 1+\varepsilon & 1 \end{bmatrix}.$$

Rescaling the first row gives

$$\begin{bmatrix} \varepsilon^{-1} & \varepsilon^{-1} \\ 1+\varepsilon & 1 \end{bmatrix}.$$

This matrix for an equivalent linear system has $det(A) = 1$ but $cond(A) = 2\varepsilon^{-1}(\varepsilon^{-1} + 1 + \varepsilon)$ which can be high for ε small.

To summarize:

- If $cond(A) = 10^t$ then at most t significant digits are lost when solving $Ax = b$.
- $cond(A) = \|A\|\|A^{-1}\|$ is the correct measure of ill-conditioning; in particular, it is scale invariant whereas $det(A)$ is not.
- For 2×2 linear systems representing two lines in the x_1, x_2 plane, $cond(A)$ is related to the angle between the lines.
- The effects of roundoff errors and finite precision arithmetic can be reduced to studying the sensitivity of the problem to perturbations.

Exercise 4.23. *Let $Ax = b$ be a square linear system and suppose you are given an approximate solution. Define the error and residual. State and prove an inequality relating the relative error, relative residual and cond(A).*

Exercise 4.24. *If A is a 2×2 matrix that is symmetric and positive definite then the $cond_2(A) = \lambda_{\max}(A)/\lambda_{\min}(A)$. If A is not symmetric there can be very little connection between the condition number and the so-called spectral condition number. Your goal in this exercise is to find an example illustrating this. Specifically, find a 2×2 matrix A with $|\lambda|_{\max}(A)/|\lambda|_{\min}(A) = O(1)$, in other words of moderate size, but $cond_2(A)$ very very large, $cond_2(A) \gg 1$. HINT: The matrix obviously cannot be symmetric. Try writing down the matrix in Jordan canonical form*

$$A = \begin{bmatrix} a & b \\ 0 & c \end{bmatrix}.$$

Exercise 4.25. *If $cond(A) = 10^5$ and one solves $Ax = b$ on a computer with 8 significant digits (base 10), what is the expected number of significant digits of accuracy in the answer? Explain how you got the result.*

Exercise 4.26. *Often it is said that "The set of invertible matrices is an open set under the matrix norm." Formulate this sentence as a mathematical theorem and prove it.*

Exercise 4.27. *For B an $N \times N$ matrix. Show that for $a > 0$ small enough then $I - aB$ is invertible. What is the infinite sum in that case:*

$$\sum_{n=0}^{\infty} a^n B^n?$$

Exercise 4.28. *Verify that the determinant gives no insight into conditioning. Calculate the determinant of the coefficient matrix of the system*

$$1x_1 + 1x_2 = 2$$
$$10.1x_1 + 10x_2 = 20.1$$

Recalculate after the first equation is multiplied by 10:

$$10x_1 + 10x_2 = 20$$
$$10.1x_1 + 10x_2 = 20.1$$

For more information see the articles and books of Wilkinson [W61], [W63].

Chapter 5

The MPP and the Curse of Dimensionality

"What we know is not much. What we do not know is immense."
— Pierre-Simon de Laplace (1749–1827) (Allegedly his last words). From: DeMorgan's Budget of Paradoxes.

5.1 Derivation

ceiiinosssttuv, ("Ut tensio, sic vis.").
— Robert Hooke

The Poisson problem is *the* model problem in mechanics and applied mathematics and the discrete Poisson problem is *the* model problem in numerical linear algebra. Since practitioners of the mad arts of numerical linear algebra will spend much of their (professional) lives solving it, it is important to understand where it comes from. Suppose "something" is being studied and its distribution is not uniform. Thus, the density of that "something" will be variable in space and possibly change with time as well. Thus, let

$$u(x,t) := \text{density}, \quad \text{where } x = (x_1, x_2, x_3).$$

For example, if something is heat, then the

$$\text{heat density} = \rho C_p T(x,t),$$

where ρ = material density, C_p = specific heat and $T(x,t)$ = temperature at point x and time t.

To avoid using "something" too much, call it Q; keep the example of heat in mind. Since Q is variable, it must be changing and hence undergoing flow with "flux" defined as its rate of flow. Thus, define

$$\overrightarrow{F} := \text{flux of } Q \text{ at a point } x \text{ at a time } t.$$

Assumption: Q is conserved.

The mathematical realization of conservation is:
For any region B

$$\frac{d}{dt}\{\text{total amount of } Q \text{ in region } B\} =$$

Total flux of Q through ∂B +

Total contribution of any sources or sinks of Q inside B.

Let, thus

$$f(x,t) := \text{ sources or sinks of } Q \text{ per unit volume.}$$

Mathematically, conservation becomes

$$\frac{d}{dt}\int_B u(x,t)dx + \int_{\partial B} \vec{F}(x,t) \cdot \hat{n}d\sigma = \int_B f(x,t)dx$$

where $\hat{n}$ is the outward unit normal to B. We shall fix x and fix the region B to be the ball about x of radius ε (denoted $B_\varepsilon(x)$). Recall the following fact from calculus about continuous functions as well as the divergence theorem.

Lemma 5.1 (Averaging of continuous functions). *If $v(x)$ is a continuous function then*

$$\lim_{\varepsilon \to 0} \frac{1}{\text{vol}(B_\varepsilon)} \int_{B_\varepsilon} v(x')dx' = v(x).$$

The divergence theorem applies in regions with smooth boundaries, with polyhedral boundaries, with rough boundaries without cusps, and many more regions. A domain must have a very exotic boundary for the divergence theorem not to hold in it. Usually, in applied math the question of "How exotic?" is sidestepped, as we do here, by just assuming the divergence theorem holds for the domain. As usual, define a "regular domain" as one to which the divergence theorem applies. We shall use if for spheres, which are certainly regular domains.

Theorem 5.1 (The Divergence Theorem). *If B is a **regular** domain (in particular, B has no internal cusps) and if $\vec{v}(x)$ is a C^1 vector function then*

$$\int_B \text{div}\,\vec{v}\,dx = \oint_{\partial B} \vec{v} \cdot \hat{n}d\sigma.$$

The divergence theorem implies

$$\int_{\partial B} \vec{F} \cdot \hat{n} d\sigma = \int_B \operatorname{div} \vec{F} dx$$

and thus (after dividing by $\operatorname{vol}(B_\varepsilon)$) conservation becomes:

$$\frac{d}{dt} \frac{1}{\operatorname{vol}(B_\varepsilon)} \int_{B_\varepsilon} u(x,t) dx + \frac{1}{\operatorname{vol}(B_\varepsilon)} \int_{B_\varepsilon} \operatorname{div}(\vec{F}) dx = \frac{1}{\operatorname{vol}(B_\varepsilon)} \int_{B_\varepsilon} f dx.$$

Letting $\varepsilon \to 0$ and using Lemma 5.1 gives the equation

$$\frac{\partial u(x,t)}{\partial t} + \operatorname{div}(\vec{F}) = f. \tag{5.1}$$

This is **one** equation for four variables (u, F_1, F_2, F_3). A connection between flux $\vec{F}$ and density is needed. One basic description of physical phenomena due to Aristotle is

"Nature abhors a vacuum".

This suggests that Q often will flow from regions of high concentration to low concentration. For example, Fourier's law of heat conduction and Newton's law of cooling state that

$$Heat\ Flux = -k\nabla T$$

where k is the (material dependent) thermal conductivity. The analogous assumption for Q is

Assumption: (Q flows downhill) *The flux in Q, is given by*

$$\vec{F} = -k\nabla u.$$

Inserting this for $\vec{F}$ in (5.1) closes the system for $u(x,t)$:

$$\frac{\partial u}{\partial t} - \operatorname{div}(k\nabla u) = f(x,t).$$

Recall that

$$\Delta u = \operatorname{div} \operatorname{grad} u = \frac{\partial^2 u}{\partial x_1^2} + \frac{\partial^2 u}{\partial x_2^2} + \frac{\partial^2 u}{\partial x_3^2}.$$

For simple materials, the value of the material parameter, k, can be taken constant. Thus,

$$\frac{\partial u}{\partial t} - k\Delta u = f(x,t).$$

If the process is at equilibrium (i.e. $u = u(x)$, independent of t, and $f = f(x)$) we have the model problem: find $u(x)$ defined on a domain Ω in $\mathbb{R}^d (d = 1, 2$ or $3)$ satisfying

$$- \Delta u = f(x) \text{ inside } \Omega, \quad u = 0 \text{ on the boundary } \partial\Omega. \qquad (5.2)$$

Remark 5.1. The boundary condition, $u = 0$ on $\partial\Omega$, is the clearest one that is interesting; it can easily be modified. Also, in $\mathbb{R}^1, \mathbb{R}^2$ and $\mathbb{R}^3$ we have

$$\Delta u = \frac{d^2}{dx^2} u(x) \text{ in } \mathbb{R}^1,$$

$$\Delta u = \frac{\partial^2 u}{\partial x_1^2} + \frac{\partial^2 u}{\partial x_2^2} \text{ in } \mathbb{R}^2,$$

$$\Delta u = \frac{\partial^2 u}{\partial x_1^2} + \frac{\partial^2 u}{\partial x_2^2} + \frac{\partial^2 u}{\partial x_3^2} \text{ in } \mathbb{R}^3.$$

Problem (5.2) is the model problem. What about the time dependent problem however? One common way to solve it is by the "method of lines" or time stepping. Pick a Δt (small) and let $u^n(x) \sim u(x,t)|_{t=n\Delta t}$.
Then

$$\frac{\partial u}{\partial t}(t_n) \doteq \frac{u^n(x) - u^{n-1}(x)}{\Delta t}.$$

Replacing $\frac{\partial u}{\partial t}$ by the difference approximation on the above RHS gives a sequence of problems

$$\frac{u^n - u^{n-1}}{\Delta t} - k\Delta u^n = f^n$$

or, solve for $n = 1, 2, 3, \cdots$,

$$-\Delta u^n + \left(\frac{1}{k}\right) u^n = f^n + u^{n-1},$$

which is a sequence of **many** shifted Poisson problems.
We shall see that:

- Solving a time dependent problem can require solving the Poisson problem (or its ilk) thousands or tens of thousands of times.
- The cost of solving the Poisson problem increases exponentially in the dimension from $1D$ to $2D$ to $3D$.

5.2   1D Model Poisson Problem

Mechanics is the paradise of the mathematical sciences, because
by means of it one comes to the fruits of mathematics.
— da Vinci, Leonardo (1452–1519), Notebooks, v. 1, ch. 20.

The $1D$ **Model Poisson Problem** (MPP henceforth) is to find $u(x)$
defined on an interval $\Omega = (a, b)$ satisfying

$$
\begin{aligned}
-u''(x) &= f(x), & a < x < b, \\
u(a) &= g_1, \quad u(b) = g_2,
\end{aligned}
\tag{5.3}
$$

where $f(x)$, g_1 and g_2 are given. If, for example, $g_1 = g_2 = 0$ the $u(x)$
describes the deflection of an elastic string weighted by a distributed load
$f(x)$. As noted in Section 5.1, $u(x)$ can also be the equilibrium temperature
distribution in a rod with external heat sources $f(x)$ and fixed temperatures
at the two ends. Although it is easy to write down the solution of (5.3), there
are many related problems that must be solved for which exact solutions
are not attainable. Thus, we shall develop method for solving all such
problems.

5.2.1   *Difference approximations*

"Finite arithmetical differences have proved remarkably suc-
cessful in dealing with differential equations, ... in this book it
is shown that similar methods can be extended to the very complicated system of differential equations which express the changes in
the weather."
— Richardson, Lewis Fry (1881–1953), page 1 from the book
Lewis F. Richardson, Weather prediction by numerical process,
Dover, New York, 1965. (originally published in 1922)

Recall from basic calculus that

$$
u'(a) = \lim_{h \to 0} \frac{u(a + h) - u(a)}{h} = \lim_{h \to 0} \frac{u(a) - u(a - h)}{h}.
$$

Thus, we can approximate by taking h nonzero but small as in:

$$
u'(a) \doteq \lim_{h \to 0} \frac{u(a + h) - u(a)}{h} =: D_+ u(a),
$$

$$
u'(a) \doteq \lim_{h \to 0} \frac{u(a) - u(a - h)}{h} =: D_- u(a).
$$

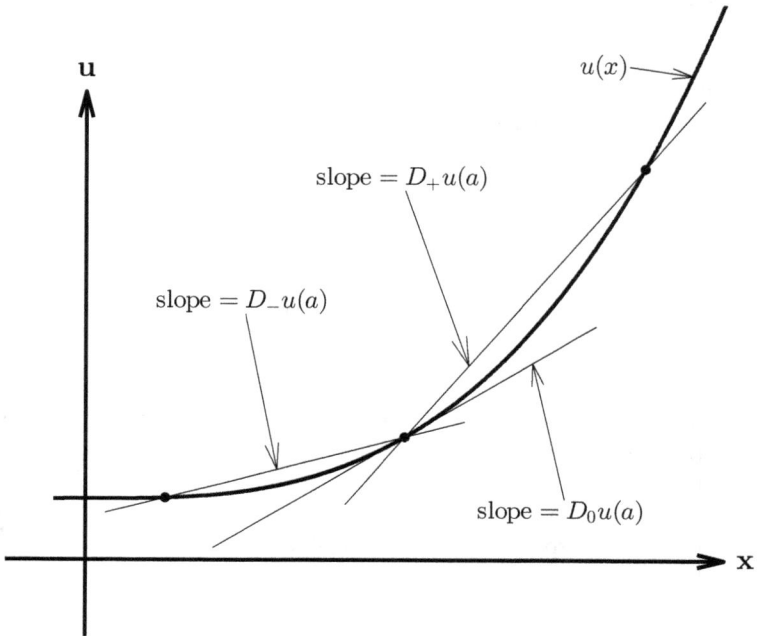

Fig. 5.1   A curve with tangent and two chords.

Graphically, we visualize these approximations as slopes of secant lines approximating the sought slopes of the tangent line as in Figure 5.1.

It seems clear that often one of $D_+u(a)$, $D_-u(a)$ will underestimate $u'(a)$ and the other overestimate $u'(a)$. Thus averaging is expected to increase accuracy (and indeed it does). Define thus

$$D_0u(a) = (D_+u(a) + D_-u(a))\,/2 = \frac{u(a+h) - u(a-h)}{2h}.$$

To reduce the model BVP to a finite set of equations, we need an approximation to u'' in (5.3). The standard one is

$$u''(a) \doteq D_+D_-u(a) = \frac{u(a+h) - 2u(a) + u(a+h)}{h^2}.$$

The accuracy of each approximation is found by using Taylor series.[1] The accuracy is (for smooth u)

$$u'(a) = D_+u(a) + O(h),$$
$$u'(a) = D_-u(a) + O(h),$$
$$u'(a) = D_0u(a) + O(h^2),$$
$$u''(a) = D_+D_-u(a) + O(h^2).$$

The expression, for example,

$$\text{error in difference approximation} = u''(a) - D_+D_-u(a) = O(h^2).$$

The expression $f(h) = O(h^2)$ means that there is a constant C so that if h is small enough, then $|f(h)| \leq Ch^2$. If h is cut in half, then $|f(h)|$ is cut by approximately a fourth; and, h is cut by 10, then $|f(h)|$ is cut by approximately 100.

5.2.2 *Reduction to linear equations*

Although this may seem a paradox, all exact science is dominated by the idea of approximation.
— Russell, Bertrand (1872–1970), in W. H. Auden and L. Kronenberger (eds.) The Viking Book of Aphorisms, New York: Viking Press, 1966.

Divide $[a, b]$ into N equal subintervals with breakpoints denoted x_j. Thus, we define

$$h := \frac{b-a}{N+1}, \quad x_j = a + jh, \quad j = 0, 1, \cdots, N+1,$$

so we have the subdivision

$$a = x_0 < x_1 < x_2 < \cdots < x_N < x_{N+1} = b.$$

At each meshpoint x_j we will compute a $u_j \sim u(x_j)$. We will, of course, need one equation for each variable meaning one equation for each meshpoint. Approximate $-u'' = f(x)$ at each x_j by using $D_+D_-u_j$:

$$-D_+D_-u_j = f(x_j),$$

or equivalently

$$-\frac{u_{j+1} - 2u_j + u_{j-1}}{h^2} = f(x_j), \quad \text{for } j = 1, 2, 3, \cdots, N.$$

---

[1]See any general numerical analysis book for this; it is not hard but would delay our presentation to take this detour.

Thus, the system of linear equations is

$$u_0 = g_0$$
$$-u_{j+1} + 2u_j - u_{j-1} = h^2 f(x_j), \quad j = 1, 2, \ldots, N,$$
$$u_{N+1} = g_1.$$

Writing this out is instructive. It is

$$
\begin{array}{lllll}
1u_0 & & & & = g_0 \\
-1u_0 & +2u_1 & -1u_2 & & = h^2 f(x_1) \\
& -1u_1 & +2u_2 & -1u_3 & = h^2 f(x_2) \\
& & \cdots & \cdots & \cdots \\
& & -1u_{N-1} & +2u_N & -u_{N+1} = h^2 f(x_N) \\
& & & & 1u_{N+1} = g_1.
\end{array}
$$

This is $N + 2$ equations in $N + 2$ variables:

$$
\begin{bmatrix}
1 & 0 & 0 & \cdots & 0 & 0 \\
-1 & 2 & -1 & \cdots & 0 & 0 \\
0 & \ddots & \ddots & \ddots & & \vdots \\
\vdots & & \ddots & \ddots & \ddots & 0 \\
0 & \cdots & 0 & -1 & 2 & -1 \\
0 & 0 & \cdots & 0 & 0 & 1
\end{bmatrix}
\begin{bmatrix}
u_0 \\
u_1 \\
\vdots \\
\vdots \\
u_N \\
u_{N+1}
\end{bmatrix}
=
\begin{bmatrix}
g_0 \\
h^2 f(x_1) \\
\vdots \\
\vdots \\
h^2 f(x_N) \\
g_1
\end{bmatrix}.
$$

The first and last equations can be eliminated (or not as you prefer) to give

$$
\begin{bmatrix}
2 & -1 & 0 & \cdots & & 0 \\
-1 & 2 & -1 & & & \\
0 & \ddots & \ddots & \ddots & & \vdots \\
\vdots & & \ddots & \ddots & \ddots & 0 \\
& & & -1 & 2 & -1 \\
0 & & \cdots & 0 & -1 & 2
\end{bmatrix}
\begin{bmatrix}
u_1 \\
u_2 \\
\vdots \\
\vdots \\
u_{N-1} \\
u_N
\end{bmatrix}
=
\begin{bmatrix}
h^2 f(x_1) + g_0 \\
h^2 f(x_2) \\
\vdots \\
\vdots \\
h^2 f(x_{N-1}) \\
h^2 f(x_N) + g_1
\end{bmatrix}.
$$

Because of its structure this matrix is often written as $A = \text{tridiag}$ $(-1, 2, -1)$. The first important question for A is:

Does this linear system have a solution?

We will investigate invertibility of A.

Lemma 5.2 (Observation about averaging). *Let a be the average of* x *and* y

$$a = \frac{x+y}{2}$$

then a must be between x and y:

    If $x < y$, then $x < a < y$,

    If $x > y$, then $x > a > y$, and

    If $x = y$ then $a = x = y$.

    More generally, the same holds for weighted averages with positive weighs: If

$$a = \alpha x + \beta y$$

where

$$\alpha + \beta = 1, \ \alpha \geq 0, \ \beta \geq 0$$

then a must be between x and y.

Exercise 5.1. *Prove this lemma about averaging.*

We will use this observation about averaging to prove that A^{-1} exists.

Theorem 5.2. *Let $A = tridiag\,(-1, 2, -1)$. Then A^{-1} exists.*

Proof. Suppose not. The $Au = 0$ has a nonzero solution u. Let u_J be the component of u that is largest in absolute value:

$$|u_J| = \max_j |u_j| \equiv u_{\text{MAX}}.$$

We can also assume $u_J > 0$; if $u_J < 0$ then note that $A(-u) = 0$ has Jth component $(-u_J) > 0$. Then, if J is not 1 or N, the Jth equation in $Au = 0$ is

$$-u_{J+1} + 2u_J - u_{J-1} = 0$$

or

$$u_J = \frac{u_{J+1} + u_{J-1}}{2}$$

implying that u_J is between u_{J-1} and u_{J+1}. Thus either they are all zero or

$$u_J = u_{J-1} = u_{J+1} \equiv u_{\text{MAX}}.$$

Continuing across the interval (a, b) we get

$$u_1 = u_2 = \ldots = u_N \equiv u_{\max}.$$

Consider the equation at x_1: $2u_1 - u_2 = 0$. Thus $2u_{\max} - 2u_{max} = 0$ so $u_{\max} \equiv 0$ and $u_J \equiv 0$. We leave the case when $J = 1$ and $J = N$ as exercises. $\qquad\square$

Exercise 5.2. *An alternative proof that the matrix given by $A =$ tridiag$(-1, 2, -1)$ is invertible involves a direct calculation of its determinant to show it is not zero. Use row-reduction operations and induction to show that $det(A) = n + 1 \neq 0$.*

Remark 5.2. The eigenvalues of the matrix $A =$ tridiag$(-1, 2, -1)$ have, remarkably, been calculated exactly. They are

$$\lambda_n = 4\sin^2 \frac{n\pi}{2N+1}, \qquad n = 1, \ldots, N.$$

Thus, $\lambda_{\max}(A) \approx 4$ and $\lambda_{\min} \approx$ Constant$\cdot h^2$, so that

$$cond_2(A) = O(h^{-2}).$$

Exercise 5.3. *Assume that $\lambda_n = 4\sin^2 \frac{n\pi}{2N+1}, n = 1, \ldots, N$. From this show that $cond_2(A) = O(h^{-2})$ and calculate the hidden constant.*

Exercise 5.4. *This exercise will calculate the eigenvalues of the 1D matrix $A =$ tridiag$(-1, 2, -1)$ exactly based on methods for solving difference equations. If $Au = \lambda u$ then, for we have the difference equation*

$$u_0 = 0,$$
$$-u_{j+1} + 2u_j - u_{j-1} = \lambda u_j, \quad j = 1, 2, \ldots, N,$$
$$u_{N+1} = 0.$$

Solutions to difference equations of this type are power functions. It is known that the exact solution to the above is

$$u_j = C_1 R_1^j + C_2 R_2^j$$

where $R_{1/2}$ are the roots of the quadratic equation

$$-R^2 + 2R - 1 = \lambda R.$$

For λ to be an eigenvalue this quadratic equation must have two real roots and there much be nonzero values of $C_{1/2}$ for which $u_0 = u_{N+1} = 0$. Now find the eigenvalues!

Exercise 5.5. *Consider the 1D convection diffusion equation (CDEqn for short): for $\varepsilon > 0$ a small number, find $u(x)$ defined on an interval $\Omega = (a, b)$ satisfying*

$$-\varepsilon u''(x) + u'(x) = f(x), \qquad a < x < b,$$
$$u(a) = g_1, \quad u(b) = g_2, \tag{5.4}$$

where $f(x)$, g_1 and g_2 are given. Let $u''(a), u'(a)$ be replaced by the difference approximations

$$u''(a) \doteq D_+D_-u(a) = \frac{u(a+h) - 2u(a) + u(a+h)}{h^2}$$

$$u'(a) \doteq D_0u(a) = \frac{u(a+h) - u(a-h)}{2h}.$$

With these approximations, the CDEqn is reduced to a linear system in the same way as the MPP. Divide $[a,b]$ into N subintervals $h := \frac{b-a}{N+1}$, $x_j = a + jh$, $j = 0, 1, \cdots, N+1$,

$$a = x_0 < x_1 < x_2 < \cdots < x_N < x_{N+1} = b.$$

At each meshpoint x_j we will compute a $u_j \sim u(x_j)$. We will, of course, need one equation for each variable meaning one equation for each meshpoint. Approximate $-u'' = f(x)$ at each x_j by using $D_+D_-u_j$:

$$-\varepsilon D_+D_-u_j + D_0u_j = f(x_j), \quad \text{for } j = 1, 2, 3, \cdots, N.$$

(a) Find the system of linear equations that results. (b) Investigate invertibility of the matrix A that results. Prove invertibility under the condition

$$Pe := \frac{h}{2\varepsilon} < 1.$$

Pe is called the cell Peclet number.

Exercise 5.6. Repeat the analysis of the 1D discrete CDEqn from the last exercise. This time use the approximation

$$u'(a) \doteq D_-u(a) = \frac{u(a) - u(a-h)}{h}.$$

5.2.3 Complexity of solving the 1D MPP

To solve the model problem we need only store a tridiagonal matrix A and the RHS b then solve $Au = b$ using tridiagonal Gaussian elimination.

Storage Costs: $\sim 4h^{-1}$ real numbers: $4 \times N = 4N$ real numbers $\sim 4h^{-1}$.
Solution Costs: $5h^{-1}$ FLOPS: $3(N-1)$ floating point operations for elimination, $2(N-1)$ floating point operations for backsubstitution.

This is a perfect result: The cost in both storage and floating point operations is proportional to the resolution sought. If we want to see the solution on scales $10\times$ finer (so $h \Leftarrow h/10$) the total costs increases by a factor of 10.

5.3   The 2D MPP

The two-dimensional model problem is the first one that reflects **some** complexities of real problems. The domain is taken to be the unit square (to simplify the problem), $\Omega = (0,1) \times (0,1)$. the problem is, given $f(x,y)$ and $g(x,y)$, to approximate the solution $u(x,y)$ of

$$
\begin{aligned}
-\Delta u &= f(x,y), \quad \text{in } \Omega, \\
u(x,y) &= g(x,y), \quad \text{on } \partial\Omega.
\end{aligned}
\tag{5.5}
$$

You can think of $u(x,y)$ as the deflection of a membrane stuck at its edges and loaded by $f(x,y)$. Figure 5.2 below given a solution where $g(x,y) = 0$ and where $f(x,y) > 0$ and so pushes up on the membrane.

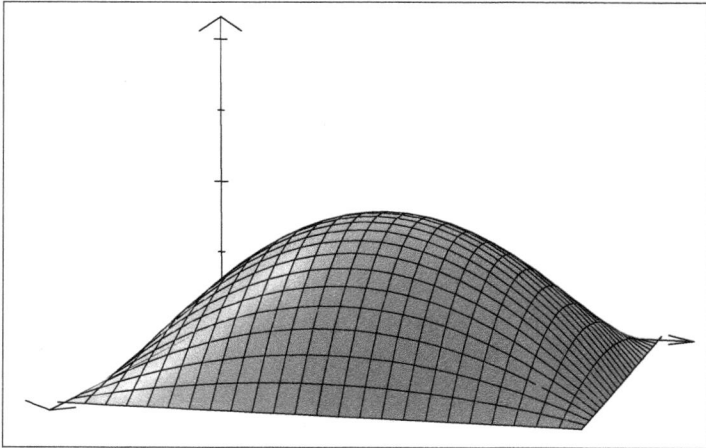

Fig. 5.2   An example solution of the 2D MPP.

Here we use (x,y) instead of (x_1, x_2) because it is more familiar, we will take $g(x,y) \equiv 0$ (to simplify the notation). Recall that

$$
-\Delta u = -(u_{xx} + u_{yy}).
$$

Different boundary conditions and more complicated domains and operators are important and interesting. However, (5.5) is the important first step to understand so we consider only (5.5) in this section.

To reduce (5.5) to a finite set of linear equations, we need to introduce a mesh and approximate u_{xx} and u_{yy} as in the $1D$ problem by their second differences in the x and y directions, respectively

$$u_{xx}(a,b) \sim \frac{u(a+\Delta x, b) - 2u(a,b) + u(a-\Delta x, b)}{\Delta x^2}, \qquad (5.6)$$

$$u_{yy}(a,b) \doteq \frac{u(a, b+\Delta y) - 2u(a,b) + u(a, b-\Delta y)}{\Delta y^2}. \qquad (5.7)$$

To use these we introduce a mesh on Ω. For simplicity, take a uniform mesh with N+1 points in both directions. Choose thus $\Delta x = \Delta y = \frac{1}{N+1} =: h$. Then set

$$x_i = ih, \quad y_j = jh, \qquad i,j = 0, 1, \ldots, N+1.$$

We let u_{ij} denote the approximation to $u(x_i, y_j)$ we will compute at each mesh point. A 10×10 mesh (h=1/10) is depicted in Figure 5.3.

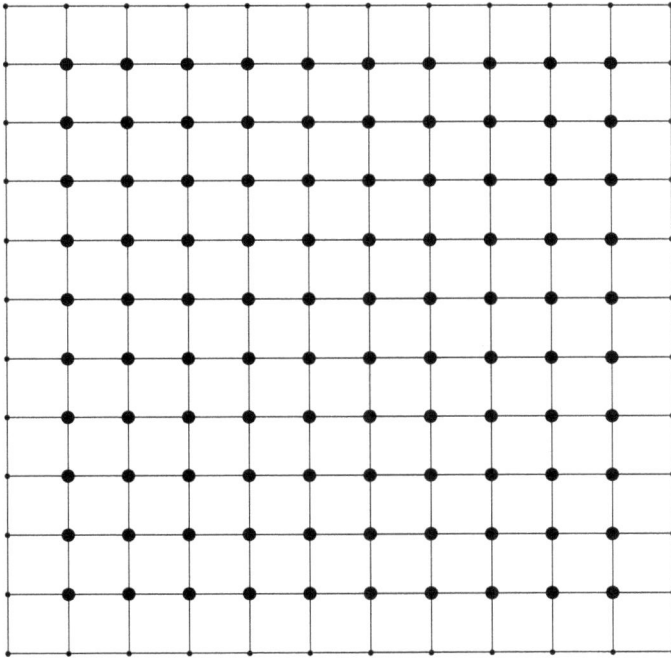

Fig. 5.3 A coarse mesh on the unit square, with interior nodes indicated by larger dots and boundary nodes by smaller ones.

To have a square linear system, we need one equation for each variable. There is one unknown (u_{ij}) at each mesh point on Ω. Thus, we need one equation at each mesh point. The equation for each mesh point on the boundary is clear:

$$u_{ij} = g(x_i, y_j) \qquad (\text{ here } g \equiv 0) \qquad \text{for each } x_i, y_j \text{ on } \partial\Omega. \qquad (5.8)$$

Thus, we need an equation for each x_i, y_j inside Ω. For a typical (x_i, y_j) inside Ω we use the approximations (5.6) and (5.7). This gives

$$-\left(\frac{u_{i+1j} - 2u_{ij} + u_{i-1j}}{h^2} + \frac{u_{ij+1} - 2u_{ij} + u_{ij-1}}{h^2} \right) = f(x_i, y_j) \qquad (5.9)$$

for all (x_i, y_j) inside of Ω.

The equations (5.8) and (5.9) give a square $(N + 2)^2 \times (N + 2)^2$ linear system for the u_{ij}'s. Before developing the system, we note that (5.9) can be simplified to read

$$-u_{i+1j} - u_{i-1j} + 4u_{ij} - u_{ij+1} - u_{ij-1} = h^2 f(x_i, y_j).$$

This is often denoted using the "difference molecule" represented by Figure 5.4,

$$-u(N) - u(S) + 4u(P) - u(E) - u(W) = h^2 f(P)$$

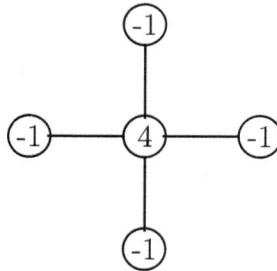

Fig. 5.4   A difference molecule with weights.

and by Figure 5.5 using the "compass" notation, where P is the mesh point and N, S, E and W are the mesh points immediately above, below, to the right and to the left of the point P.

The equation, rewritten in terms of the stencil notation, becomes

$$-u(N) - u(S) + 4u(C) - u(E) - u(W) = h^2 f(C).$$

The *discrete* Laplacian, denoted Δ^h, in 2D is thus

$$-\Delta^h u_{ij} := \frac{-u_{ij+1} - u_{ij-1} + 4u_{ij} - u_{i+1j} - u_{i-1j}}{h^2},$$

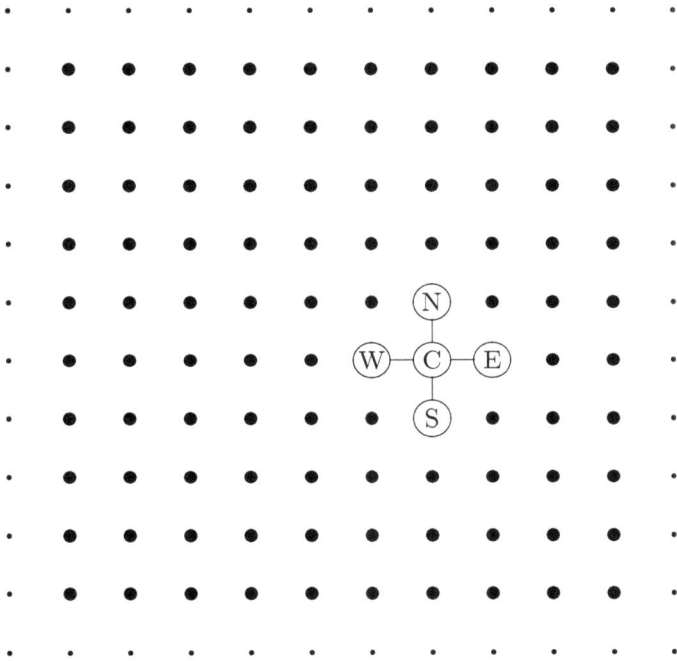

Fig. 5.5 A sample mesh showing interior points and indicating a five-point Poisson equation stencil and the "compass" notation, where C is the mesh point and N, S, E, W are the mesh points immediately above, below, to the right and the left of C.

so the equations can be written compactly as

$$-\Delta^h u_{ij} = f(x_i, y_j), \quad \text{at all } (x_i, y_j) \text{ inside } \Omega,$$
$$u_{ij} = g(x_i, y_j) \ (\equiv 0 \ ) \text{ at all } \ (x_i, y_j) \ \text{ on } \ \partial\Omega. \tag{5.10}$$

The boundary unknowns can be eliminated so (5.10) becomes an $N^2 \times N^2$ linear system for the N^2 unknowns:

$$A_{N^2 \times N^2} \, u_{N^2 \times 1} = f_{N^2 \times 1}. \tag{5.11}$$

Since each equation couples u_{ij} to its four nearest neighbors in the mesh, A will typically have only 5 nonzero entries per row. To actually find A we must order the unknowns u_{ij} into a vector u_k, $k = 1, 2, \ldots, N^2$. A lexicographic ordering is depicted in Figure 5.6.

Thus, through the difference stencil, if u_{ij} is the kth entry in u, u_{ij} is linked to u_{k-1}, u_{k+1}, u_{k+N} and u_{k-N}, as in Figure 5.7.

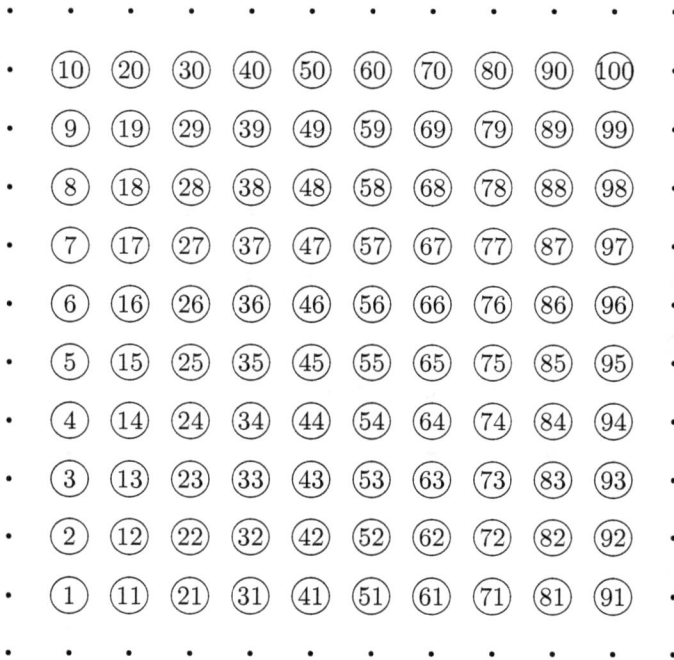

Fig. 5.6   Node numbering in a lexicographic order for a 10 × 10 mesh.

Thus, the *typical* k^{th} row (associated with an *interior* mesh point (x_i, y_j) not adjacent to a boundary point) of the matrix A will read:

$$(0, 0, \ldots, 0, -1, 0, 0, \ldots, 0, -1, 4, -1, 0, \ldots, 0, -1, 0, \ldots, 0). \qquad (5.12)$$

Conclusion: A is an $N^2 \times N^2$ ($\simeq h^{-2} \times h^{-2}$) sparse, banded matrix. It will *typically* have only 5 nonzero entries per row. Its bandwidth is $2N + 1$ and it's half bandwidth is thus $p = N (\simeq h^{-1})$.

Complexity Estimates: For a given resolution h or given N storing A as a banded matrix requires storing

$$2(N + 1) \times N^2 \text{ real numbers} \simeq 2h^{-3} \text{ real numbers.}$$

Solving $Au = f$ by banded Gaussian elimination requires

$$O((2N + 1)^2 N^2) = O(N^4) \simeq O(h^{-4}) \text{ FLOPS.}$$

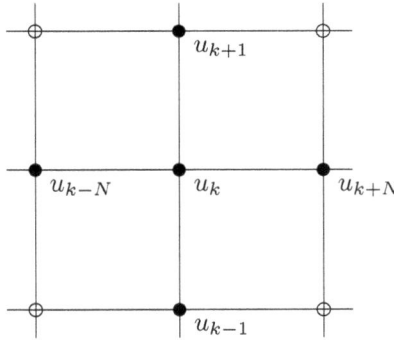

Fig. 5.7 The 2D difference stencil at the kth point in a lexicographic ordering with N numbered points in each direction.

The exact structure of A can easily be written down because all the choices were made to keep A as simple as possible. A is an $N^2 \times N^2$ block tridiagonal matrix ($N \times N$ blocks with each block an $N \times N$ matrix) of the form:

$$A = \begin{bmatrix} T & -I & & 0 \\ -I & T & \ddots & \\ & \ddots & \ddots & -I \\ 0 & & -I & T \end{bmatrix} \tag{5.13}$$

where I is the $N \times N$ identity matrix and

$$T = tridiag(-1, 4, -1) \quad (N \times N \text{ matrix}).$$

Exercise 5.7. *Consider the 2D MPP with RHS $f(x, y) = x - 2y$ and boundary condition $g(x, y) = x - y$. Take $h = 1/2$ and write down the difference approximation to $u(1/2, 1/2)$. Compute is value.*

Exercise 5.8. *Consider the 2D MPP with RHS $f(x, y) = x - 2y$ and boundary condition $g(x, y) = x - y$. Take $h = 1/3$ and write down the 4×4 linear system for the unknown values of u_{ij}.*

Exercise 5.9. *The N^2 eigenvalues and eigenvectors of the matrix A in (5.13) have been calculated exactly, just as in the 1D case. They are, for $n, m = 1, \ldots, N$,*

$$\lambda_{n,m} = 4 \left(\sin^2 \frac{n\pi}{2(N+1)} + \sin^2 \frac{m\pi}{2(N+1)} \right),$$

and

$$(\vec{u}_{n,m})_{j,k} = \sin\frac{jn\pi}{N+1}\sin\frac{km\pi}{N+1},$$

where j and k vary from $1,\ldots,N$. *Verify these expressions by calculating* $A\vec{u}_{n,m}$ *and showing it is equal to* $\lambda_{n,m}\vec{u}_{n,m}$.

Exercise 5.10. *Let the domain be the triangle with vertices at* $(0,0)$, $(1,0)$, *and* $(0,1)$. *Write down the linear system arising from the MPP on this domain with* $f(x,y) = x + y, g = 0$ *and* $N = 5$.

5.4   The 3D MPP

The 3D model Poisson problem is to find

$$u = u(x,y,z)$$

defined for (x,y,z) in the unit cube

$$\Omega := \{(x,y,z)|0 < x,y,z < 1\}$$

satisfying

$$-\Delta u = f(x,y,z) \ , \ \text{in} \ \ \Omega,$$

$$u = g(x,y,z), \ \text{on the boundary} \ \partial\Omega.$$

The Laplace operator in $3D$ is (writing it out)

$$\Delta u = \text{div grad } u = \frac{\partial^2 u}{\partial x^2} + \frac{\partial^2 u}{\partial y^2} + \frac{\partial^2 u}{\partial z^2},$$

and a discrete Laplacian is obtained by approximating each term by the 1D difference in the x, y and z directions. We shall now develop the linear system arising from the usual central difference model of this problem, making the simplest choice at each step. First, take

$$h = \frac{1}{N+1}, \quad \text{and set}$$

$$\Delta x = \Delta y = \Delta z = h = \frac{1}{N+1}.$$

Define the mesh points in the cube as

$$x_i = ih, y_j = jh, z_k = kh \text{ for } 0 \le i,j,k \le N+1.$$

Thus a typical mesh point is the triple (x_i, y_j, z_k). There are $(N+2)^3$ of these but some are known from boundary values; there are exactly N^3 of these that need to be calculated. Thus we must have an $N^3 \times N^3$ linear system: one equation for each unknown variable! Let the approximation at the meshpoint (x_i, y_j, z_k) be denoted (as usual) by

$$u_{ijk} := \text{approximation to } u(x_i, y_j, z_k).$$

The discrete Laplacian in $3D$ is

$$\Delta^h u_{ijk} := \frac{u_{i+1jk} - 2u_{ijk} + u_{i-1jk}}{h^2}$$
$$+ \frac{u_{ij+1k} - 2u_{ijk} + u_{ij-1k}}{h^2} + \frac{u_{ijk+1} - 2u_{ijk} + u_{ijk-1}}{h^2}.$$

Collecting terms we get

$$\Delta^h u_{ijk} :=$$
$$\frac{u_{i+1jk} + u_{ij+1k} + u_{ijk+1} - 6u_{ijk} + u_{i-1jk} + u_{ij-1k} + u_{ijk-1}}{h^2}.$$

The $3D$ discrete model Poisson problem is thus

$$-\Delta^h u_{ijk} = f(x_i, y_j, z_k), \text{ at all meshpoints } (x_i, y_j, z_k) \text{ inside } \Omega$$

$$u_{ijk} = g(x_i, y_j, z_k) = 0, \text{ at all meshpoints } (x_i, y_j, z_k) \text{ on } \partial\Omega.$$

In the above "*at all meshpoints (x_i, y_j, z_k) inside Ω*"means for $1 \le i, j, k \le N$ and "*at all meshpoints (x_i, y_j, z_k) on $\partial\Omega$*" means for i or j or $k = 0$ or $N + 1$. We thus have the following square (one variable for each meshpoint and one equation at each meshpoint) system of linear equations (where $f_{ijk} := f(x_i, y_j, z_k)$).
For $1 \le i, j, k \le N$,

$$-u_{i+1jk} - u_{ij+1k} - u_{ijk+1} + 6u_{ijk} - u_{i-1jk} - u_{ij-1k} - u_{ijk-1} = h^2 f_{ijk}.$$

And for i or j or $k = 0$ or $N + 1$,

$$u_{ijk} = 0. \tag{5.14}$$

The associated difference stencil is sketched in Figure 5.8.

Counting is good!

This system has one unknown per meshpoint and one equation per meshpoint. In this form it is a square $(N + 2)^3 \times (N + 2)^3$ linear system. Since $u_{ijk} = 0$ for all boundary meshpoints we can also eliminate these degrees of freedom and get a reduced[2] $N^3 \times N^3$ linear system, for $1 \le i, j, k \le N$:

$$-u_{i+1jk} - u_{ij+1k} - u_{ijk+1} + 6u_{ijk} - u_{i-1jk} - u_{ij-1k} - u_{ijk-1} = h^2 f_{ijk},$$

---

[2]We shall do this reduction herein. However, there are serious reasons not to do it if you are solving more general problems: including these gives a negligably smaller system and it is easy to change the boundary conditions. If one eliminates these unknowns, then changing the boundary conditions can mean reformatting all the matrices and programming again from scratch. On the other hand, this reduction results in a symmetric matrix while keeping Dirichlet boundary conditions in the matrix destroys symmetry and complicates the solution method.

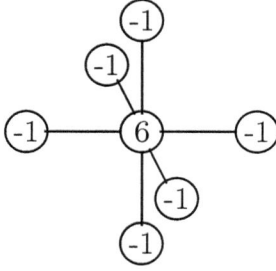

Fig. 5.8   The difference molecule or stencil in 3D.

where subscripts of i, j, or k equal to 0 or $N + 1$ are taken to mean that $u_{ijk} = 0$.

The complexities of these connections is revealed by considering the nearest neighbors on the physical mesh that are linked in the system. A uniform mesh is depicted in Figures 5.9 and 5.10.

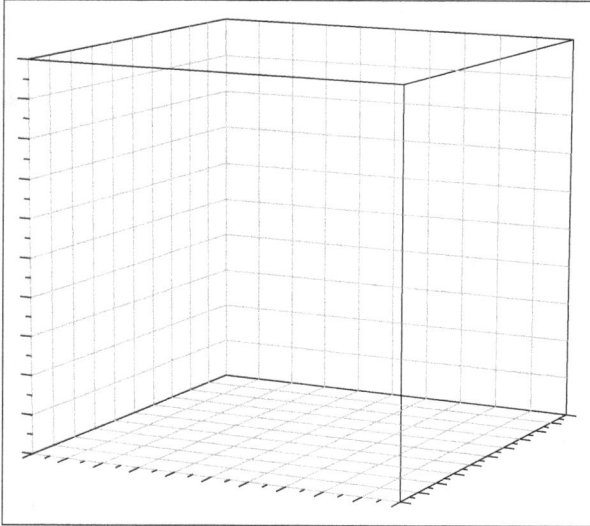

Fig. 5.9   A 3D mesh.

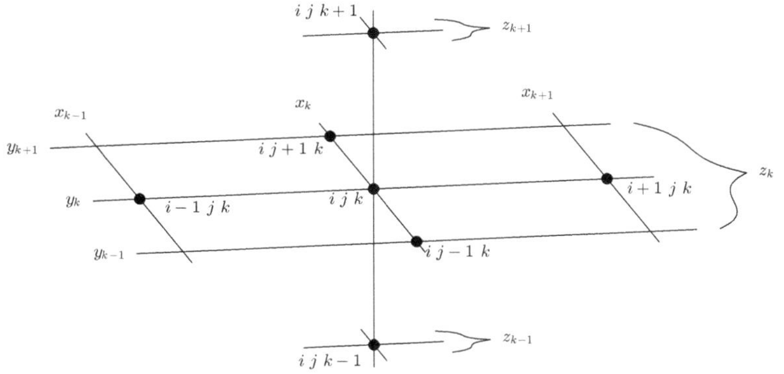

Fig. 5.10 Geometry of a 3D uniform mesh. Each point has six neighbors, indicated by heavy dots.

A *typical row* in the matrix (when a lexicographic ordering of meshpoints is used) looks like

$$0,\ldots,0,-1,\underbrace{0,\ldots\ldots,0}_{N^2-N-1},-1,\underbrace{0,\ldots 0}_{N-1},-1,\quad 6,$$

$$-1,\underbrace{0,\ldots,0}_{N-1},-1,\underbrace{0,\ldots\ldots,0}_{N^2-N-1},-1,0,\ldots,0$$

where the value 6 is the diagonal entry. If the mesh point is adjacent to the boundary then this row is modified. (In 3D adjacency happens often.)

To summarize, some basic facts about the coefficient matrix A of linear system derived from the 3D model Poisson problem on an $N \times N \times N$ mesh with $h = 1/(N+1)$:

- A is $N^3 \times N^3$ (huge if $N = 100$ say!).
- A has at most 7 nonzero entries in each row.
- N^3 equations with *bandwidth* $= 2N^2 + 1$ or half bandwidth $p = N^2$.
- Storage as a banded sparse matrix requires storing
$$N^3 \times (2N^2 + 1) \doteq 2N^5 \text{ real numbers.}$$
- Solution using banded sparse Gaussian elimination requires about
$$O((2N^2 + 1)^2 \times N^3) = O(N^7) \text{ FLOPS.}$$
- Suppose you need 10× more resolution (so $\Delta x \leftarrow \Delta x/10$, $\Delta y \leftarrow \Delta y/10$ and $\Delta z \leftarrow \Delta z/10$). Then $h \to h/10$ and thus $N \to 10N$. It follows

that

> Storage requirements increase $100,000$ times, and
>
> Solution by banded sparse GE takes $10,000,000$ longer!

The 3D matrix has certain mathematical similarities to the 1D and 2D matrices. Exploiting these one can show.

Theorem 5.3 (Eigenvalues of the discrete MPP). *Let A denote the coefficient matrix arising from the 3D model Poisson problem on a uniform mesh with $h = 1/(N+1)$. Then A is nonsingular. Furthermore, the eigenvalues of A are given by*

$$\lambda_{pqr} = 4 \left(\sin^2 \left(\frac{p\pi}{2(N+1)} \right) + \sin^2 \left(\frac{q\pi}{2(N+1)} \right) \right.$$

$$\left. + \sin^2 \left(\frac{r\pi}{2(N+1)} \right) \right), \quad for \quad 1 \le p, q, r \le N.$$

Thus,

$$\lambda_{\max}(A) \doteq 12, \lambda_{\min}(A) \doteq h^2,$$

$$cond_2(A) = O(h^{-2}).$$

Exercise 5.11. *Prove the claimed estimates of $cond(A)$ from the formula for the eigenvalues of A.*

5.5   The Curse of Dimensionality

> "Let us first understand the facts, and then we may seek for the causes."
>
> — Aristotle.

The right way to compare the costs in storage and computer time of solving a BVP is in terms of the resolution desired, i.e., in terms of the meshwidth h. The previous estimates for storage and solution are summarized in Table 5.1. Comparing these we see the curse of dimensionality clearly: As the dimension increases, the exponent increases rather than the constant or parameter being raised to the exponent. In other words:

The cost of storing the data and solving the linear system using direct methods for the model problem increases exponentially with the dimension.

Table 5.1   Costs of banded sparse GE.

	1D	2D	3D
Storage cost (# real numbers)	$4h^{-1}$	$2h^{-3}$	$2h^{-5}$
Solution cost (# FLOPs)	$5h^{-1}$	$O(h^{-4})$	$O(h^{-7})$

To put some concrete numbers to this observation, on a typical inexpensive PC in 2012, one is able to store the data for a $2D$ model problem on a mesh with $h \simeq 1/500$. In other words, one can store roughly

$$\text{in } 2D: \quad 2 \cdot 500^3 = 250,000,000 \quad \text{double precision numbers.}$$

If one is solving the $1D$ problem instead with this computer it could store a matrix with $h = h_{\min}$ where

$$4h_{\min}^{-1} = 250,000,000, \quad \text{or} \quad h_{\min} = 1.6 \times 10^{-8}$$

which is an exceedingly small meshwidth. On the other hand, if you were solving a $3D$ problem, the finest mesh you can store is

$$2h_{\min}^{-5} = 250,000,000, \quad \text{or} \quad h_{\min} \simeq \frac{1}{40},$$

which is exceedingly coarse.

Using the same kind of estimates, suppose storage is not an issue and that for $h = 1/1000$ solving the $2D$ problem takes 100 minutes. This means the completely hypothetical computer is doing roughly 10^{-10} minute/flop. From the above table we would expect the time required to solve the $1D$ and $3D$ problems for the same resolution, $h = 1/1000$, to be:

in 1D: 10^{-7} minutes,
in 3D: 10^{+11} minutes!

This is the curse of dimensionality in turnaround times. In practical settings, often programs are used for design purposes (solve, tweak one design or input parameter, solve again, see what changes) so to be useful one needs at least 1 run per day and 3 runs per day are desired.

How is one to break the curse of dimensionality? We start with one key observation.

5.5.1   *Computing a residual grows slowly with dimension*

Though this be madness, yet there is method in't.
— Shakespeare, William (1564–1616)

Given an approximate solution to the model Poisson problem we can compute a residual cheaply since A only has a few nonzero entries per row.

The 1D case: We have[3]

```
for i=1:N
  r(i) = h^2*f(i)-(-u(i+1)+2u(i)-u(i-1))
end
```

This takes 3 multiplications and 3 additions per row giving $6h^{-1}$ FLOPS. Notice that the matrix does not need to be stored—only the vectors **f**, **u** and **r**, of length $N \simeq 1/h$.

The 2D case: The $2D$ case takes 5 multiplies and 5 adds per row for h^{-2} rows by:

```
for i=1:N
  for j=1:N
    r(i,j)=h^2*f(i,j)-(-u(i,j+1)-u(i,j-1) + 4u(i,j)
                       -u(i+1,j)-u(i-1,j) )
  end
end
```

This gives a total of only $10h^{-2}$ FLOPS and requires only $2h^{-2}$ real numbers to be stored.

The 3D case: The $3D$ case takes 7 multiplies and adds per row for h^{-3} rows by:

```
for i=1:N
  for j=1:N
    for k=1:N
      r(i,j,k)=h^2*f(i,j,k) - ( ...
                          -u(i+1,j,k)-u(i,j+1,k)-u(i,j,k+1)
                          +6u(i,j) ...
                          -u(i-1,j,k)-u(i,j-1,k)-u(i,j,k-1) )
    end
  end
end
```

This gives a total of only $14h^{-3}$ FLOPS and requires only $2h^{-3}$ real numbers to be stored.

---

[3]When i=1, the expression "u(i-1) " is to be interpreted as the boundary value at the left boundary, and when i=N, the expression "u(i+1)" is to be interpreted as the boundary value at the right boundary.

To summarize,

Costs for computing residual			
dimension of model	1D	2D	3D
# real numbers storage	$2h^{-1}$	$2h^{-2}$	$2h^{-3}$
# FLOPS	$6h^{-1}$	$10h^{-2}$	$14h^{-3}$

The matrix A does not need to be stored for the MPP since we already know the nonzero values and the components they multiply. More generally we would only need to store the nonzero entries and a pointer vector to tell which entry in the matrix is to be multiplied by that value. Thus the only hope to break the curse of dimensionality is to use algorithms where the work involves computing residuals instead of elimination! These special methods are considered in the next chapter.

Exercise 5.12. *Write a program to create, as an array, the matrix which arises from the $2D$ model Poisson problem on a uniform $N \times N$ mesh. Start small and increase the dimension (N and the dimension of the matrices).*

(1) *Find the smallest h (largest N) for which the program will execute without running out of memory.*
(2) *Next from this estimate and explain how you did it: the smallest h (largest N) for which this can be done in $2D$ in banded sparse storage mode. The same question in $1D$. The same question in $3D$.*
(3) *Make a chart of your findings and draw conclusions.*

Exercise 5.13. *If solving a $2D$ MPP program takes 30 minutes with $N = 20000$, estimate how long it would take to solve the problem with the same value of h in $1D$ and in $3D$. Explain.*

Exercise 5.14. *Same setting as the last problem. Now however, estimate how long it would take to compute a residual in $1D, 2D$ and $3D$. Explain how you did the estimate.*

Exercise 5.15. *Think about the problem of computing Ax where A is large and sparse but with a non zero structure less regular than for the MPP. Thus, the non zero entries in A must be stored as well as (for each) somehow the row and column number in which that entry appears. Formalize one way to store A in this manner then write down in pseudo code how to compute $x \rightarrow Ax$. Many people have worked on sparse matrix storage schemes so it is unlikely that your solution will be best possible. However,* after finding one answer, you will be able to quickly grasp the point of the various sparse storage schemes. Next look in the Templates book, Barrett, Berry, et al. [Barrett et al. (1994)] and compare your method to Compressed Row Storage. Explain the differences.

Chapter 6

Iterative Methods

"The road to wisdom? Well, it is plain
And simple to express:
Err and err and err again,
But less and less and less."
— Piet Hein[1]

6.1 Introduction to Iterative Methods

Iterative methods for solving $Ax = b$ are rapidly becoming the workhorses of *parallel and large scale* computational mathematics. Unlike Gaussian elimination, using them reliably depends on knowledge of the methods and the matrix and there is a wide difference in the performance of different methods on different problems. They are particularly important, and often the only option, for problems where A is large and sparse. The key observation for these is that to compute a matrix-vector multiply, $\widehat{x} \to A\widehat{x}$, one only needs to store the nonzero entries in A_{ij} and their indices i and j. These are typically stored in some compact data structure that does not need space for the zero entries in A. If the nonzero structure of A is regular, as for the model Poisson problem on a uniform mesh, even i and j need not be stored!

Consider the problem of solving linear systems

$$Ax = b, \qquad A : \text{large and sparse.}$$

As in chapter 5, computing the residual

$$r = b - A\widehat{x}, \qquad \widehat{x} : \text{an approximate solution}$$

---

[1]Piet Hein is a famous Danish mathematician, writer and designer. He is perhaps most famous as a designer. Interestingly, he is a descendent of another Piet Hein from the Netherlands who is yet more famous in that small children there still sing songs praising him.

is cheap in both operations and storage. Iterative methods take a form exploiting this, generally resembling:

Algorithm 6.1 (Basic iterative method). *Given an approximate solution* $\widehat{x}$ *and a maximum number of steps* itmax*:*

Compute residual: $\widehat{r} = b - A\widehat{x}$
for i = 1:itmax
    Use $\widehat{r}$ *to improve* $\widehat{x}$
    Compute residual using improved $\widehat{x}$: $\widehat{r} = b - A\widehat{x}$
    Use residual and update to estimate accuracy
    if *accuracy is acceptable, exit with converged solution*
end
Signal failure if accuracy is not acceptable.

As an example (that we will analyze in the next section) consider the method known as first order Richardson, or FOR. In FOR, we pick the number $\rho > 0$, rewrite $Ax = b$ as

$$\rho(x - x) = b - Ax,$$

then guess x^0 and iterate using

$$\rho(x^{n+1} - x^n) = b - Ax^n, \text{ or}$$
$$x^{n+1} = \left[I - \frac{1}{\rho}A\right]x^n + \frac{1}{\rho}b.$$

Algorithm 6.2 (FOR = First Order Richardson). *Given* $\rho > 0$, *target accuracy* tol, *maximum number of steps* itmax *and initial guess* x^0:

Compute residual: $r^0 = b - Ax^0$
for n = 1:itmax
    Compute update $\Delta^n = (1/\rho)r^n$
    Compute next approximation $x^{n+1} = x^n + \Delta^n$
    Compute residual $r^{n+1} = b - Ax^{n+1}$
    Estimate residual accuracy criterion $\|r^{n+1}\|/\|b\|$ <tol
    Estimate update accuracy criterion $\|\Delta^n\|/\|x^{n+1}\|$ <tol
    if *both residual and update are acceptable*
        exit with converged solution
    end
end
Signal failure if accuracy is not acceptable.

We shall see that FOR is a terrific iterative method for introducing the ideas and mathematics of the area but a very slow one for actually solving $Ax = b$. Nevertheless, if there were no faster ones available, then it would still be very widely used because of the curse of dimensionality. To understand why, let us return to the example of the model Poisson problem in 3D discussed in the previous chapter. Recall that, for a typical point $(x_i, y_j, z_k) \in \Omega$, the equation becomes

$$6u_{ijk} - u_{i+1jk} - u_{i-1jk} - u_{ij+1k} - u_{ij-1k} - u_{ijk+1} - u_{ijk-1} = h^2 f_{ijk},$$

where $u_{ijk} = u(x_i, y_j, z_k)$ and $f_{ijk} = f(x_i, y_j, z_k)$ and if any point lies on the boundary, its value is set to zero. Picking $\rho = 6$ (called the Jacobi method) makes FOR particularly simple, and it is given by

Algorithm 6.3 (Jacobi iteration in 3D). *Given a tolerance* tol, *a maximum number of iterations* itmax *and arrays* uold, unew *and* f, *each of size* (N+1,N+1,N+1). *with boundary values[2] of* uold *and* unew *filled with zeros:*

```
h=1/N
for it=1:itmax
  % initialize solution, delta, residual and rhs norms
  delta=0
  unorm=0
  resid=0
  bnorm=0

  for i=2:N
    for j=2:N
      for k=2:N
        % compute increment
        au=-( uold(i+1,j,k) + uold(i,j+1,k) ...
            + uold(i,j,k+1) + uold(i-1,j,k) ...
            + uold(i,j-1,k) + uold(i,j,k-1) )
        unew(i,j,k)=(h^2*f(i,j,k) - au)/6

        % add next term to norms
        delta=delta + (unew(i,j,k) - uold(i,j,k))^2
        unorm=unorm + (unew(i,j,k))^2
        resid=resid + (h^2*f(i,j,k) ...
```

---

[2]Boundaries are locations for which i or j or k take on the values 1 or (N+1).

```
                                    - au - 6*uold(i,j,k))^2
            bnorm=bnorm + (h^2*f(i,j,k))^2
          end
      end
  end


  uold=unew  % set uold for next iteration

  % complete norm calculation
  delta=sqrt(delta)
  unorm=sqrt(unorm)
  resid=sqrt(resid)
  bnorm=sqrt(bnorm)

  % test for convergence
  if resid<tol*bnorm & delta<tol*unorm
    'solution converged'
    return
  end

end
error('convergence failed')
```

Remark 6.1. If Algorithm 6.3 were written to be executed on a computer, the calculation of **bnorm** would be done *once*, before the loop began. Calculating it on each iteration is a waste of computer time because it never changes.

Programming the Jacobi method in this way is particularly simple for a uniform mesh (as above). The computations reflect the underlying mesh. Each approximate solution is computed at a given mesh point by averaging the values of the points 6 nearest neighbors then adding this to the right hand side. This style of programming is obviously parallel (think of 1 CPU at each point on the mesh with nearest neighbor connections). Unfortunately, the program must be rewritten from scratch whenever the geometry of the mesh connectivity changes.

The Jacobi method requires that only three $N \times N \times N$ arrays be stored: $\mathtt{f(i,j,k)}$, containing the value $f(x_i, y_j, z_k)$, and $\mathtt{uold(i,j,k)}$ and $\mathtt{unew(i,j,k)}$ containing the values of the old and new (or updated) approximations. Remarkably, this does not require that the coefficient matrix be stored at all! Thus, *provided it converges rapidly enough*, we have a method for overcoming the curse of dimensionality. Unfortunately, this *"provided"* is the key question: Iterative methods utility depend on speed of convergence and, double unfortunately, we shall see that the Jacobi method does not converge fast enough as the next example begins to indicate.

Example 6.1 (FOR for the $1D$ model Poisson problem). The $1D$ model problem

$$-u'' = f(x), \qquad 0 < x < 1, u(0) = 0 = u(1)$$

with $h = \frac{1}{5}$ and $f(x) = \frac{x}{5}$ leads to the 4×4 tridiagonal linear system

$$\begin{array}{rrrr}
2u_1 & -u_2 & & = 1 \\
-u_1 & +2u_2 & -u_3 & = 2 \\
& -u_2 & +2u_3 & -u_4 = 3 \\
& & -u_3 & +2u_4 = 4.
\end{array}$$

The true solution is

$$u_1 = 4, \quad u_2 = 7, \quad u_3 = 8, \quad u_4 = 6.$$

Taking $\rho = 2$ in FOR gives the iteration

$$u_1^{\text{NEW}} = \frac{1}{2} + \frac{1}{2}u_2^{\text{OLD}}$$

$$u_2^{\text{NEW}} = \frac{2}{2} + (u_1^{\text{OLD}} + u_3^{\text{OLD}})/2$$

$$u_3^{\text{NEW}} = \frac{3}{2} + (u_2^{\text{OLD}} + u_4^{\text{OLD}})/2$$

$$u_4^{\text{NEW}} = \frac{4}{2} + u_3^{\text{OLD}}/2.$$

(This is also known as the Jacobi iteration.) Taking $u^0 = u^{\text{OLD}} = (0,0,0,0)^t$ we easily compute the iterates

$$u^1 = \begin{bmatrix} 1/2 \\ 1 \\ 3/2 \\ 2 \end{bmatrix}, \; u^{10} = \begin{bmatrix} 3.44 \\ 6.07 \\ 7.09 \\ 5.42 \end{bmatrix}, \; u^{20} = \begin{bmatrix} 3.93 \\ 6.89 \\ 7.89 \\ 5.93 \end{bmatrix}, \; u^{35} = \begin{bmatrix} 4.00 \\ 7.00 \\ 8.00 \\ 6.00 \end{bmatrix}.$$

This problem is only a 4×4 linear system and can be very quickly solved exactly by hand. To solve to 2 digits by the Jacobi method took 35 steps which is much slower.

Exercise 6.1. *For the choices below do 2 steps of FOR*

$$\rho = 2, A = \begin{bmatrix} 2 & 1 \\ 1 & 2 \end{bmatrix}, b = \begin{bmatrix} 1 \\ -1 \end{bmatrix}, x^0 = \begin{bmatrix} 0 \\ 0 \end{bmatrix}.$$

Exercise 6.2.

(1) *Write a computer program to apply FOR with $\rho = 2$ (Jacobi iteration) to the 1D Model Poisson Problem as described in Example 6.1, performing only a fixed number of iterations, not checking for convergence. Check your work by verifying the four iterates given in Example 6.1.* **Warning:** *The four values u_1, u_2, u_3 and u_4 in Example 6.1 would refer to the four values* u(2), u(3), u(4) *and* u(5) *if Algorithm 6.3 were to be written for 1D. This is because* u(1)=0 *and* u(6)=0.

(2) *Add convergence criteria as described in Algorithm 6.3 to your code. For the same problem you did in part 1, how many iterations would be required to attain a convergence tolerance of 10^{-8}?*

(3) *For the 1D Model Poisson Problem that you coded above in part 1, but with $h = 1/200$ and $f(x) = x/200$, how many iterations would be required to attain a convergence tolerance of 10^{-8}?*

(4) *It is easy to find the analytic solution the 1D Model Poisson Problem by integrating twice. Compare the analytic solution with your computed solution with $h = 1/200$ by computing the relative difference between the two at each of the 200 mesh points and finding the square root of the sum of squares of these differences divided by the square root of the sum of squares of either solution (the relative two norm).*

Exercise 6.3. *Consider the 2D MPP with RHS $f(x, y) = x - 2y$ and boundary condition $g(x, y) = x - y$. Take $h = 1/3$ and write down the 4×4 linear system for the unknown values of u_{ij}. Take $\rho = 2$ and initial guess the zero vector and do 2 steps of FOR. It will be easier to draw a big picture of the physical mesh and do the calculations on the picture than to write it all out as a matrix.*

Exercise 6.4. *Consider the 2D MPP with RHS $f(x, y) = 0$ and with boundary condition $g(x, y) = x - y$.*

(1) *Write a computer program to solve this problem for $h = 1/N$ using FOR with $\rho = 4$ (Jacobi iteration). This amounts to modifying Algorithm 6.3 for 2D instead of 3D.*

(2) *It is easy to see that the solution to this problem is $u(x, y) = x - y$. Remarkably, this continuous solution is also the discrete solution.*

Verify that your code reproduces the continuous solution to within the convergence tolerance for the case $h = 1/3$ $(N = 3)$.

(3) Verify that your code reproduces the continuous solution to within the convergence tolerance for the case $h = 1/100$ $(N = 100)$.

6.1.1 *Iterative methods three standard forms*

"Truth emerges more readily from error than from confusion."
— Francis Bacon

FOR can be generalized by replacing the value ρ with a matrix M. It can be written as

Algorithm 6.4 (Stationary iterative method). *Given a $N \times N$ matrix A, another $N \times N$ matrix M, a right side vector b and an initial guess x^0,*

```
n=0
while convergence is not satisfied
    Obtain x^{n+1} as the solution of M(x^{n+1} − x^n) = b − Ax^n
    n=n+1
end
```

The matrix M does not depend on the iteration counter n, hence the name "stationary." This algorithm results in a new iterative method for each new choice of M, called a "preconditioner." For FOR (which takes very many steps to converge) $M = (1/\rho)I$. At the other extreme, if we pick $M = A$ then the method converges in 1 step but that one step is just solving a linear system with A so no simplification is obtained. From these two extreme examples, it is expected that some balance must be struck between the cost per step (less with simpler M) and the number of steps (the closer M is to A the fewer steps expected).

Definition 6.1. Given an $N \times N$ matrix A, an $N \times N$ matrix M that approximates A in some useful sense and for which the linear system $My = d$ is easy to solve is a **preconditioner** of A.

Definition 6.2. For a function Φ, a **fixed point** of $\Phi(x)$ is any x satisfying $x = \Phi(x)$ and a **fixed point iteration** is an algorithm approximating x by guessing x^0 and repeating $x^{n+1} = \Phi(x^n)$ until convergence.

There are three standard ways to write any stationary iterative method.

(1) **Residual-Update Form:** $r^n = b - Ax^n$ is the residual and $\Delta^n = x^{n+1} - x^n$ is the update. Thus, the residual-update form is: given x^n,

$$r^n = b - Ax^n$$
$$\Delta^n = M^{-1}r^n$$
$$x^{n+1} = x^n + \Delta^n.$$

This is often the way the methods are programmed.

(2) **Fixed Point Iteration Form:** A stationary iterative method can be easily rewritten as a fixed point iteration. Define $T = I - M^{-1}A$ then we have

$$x^{n+1} = M^{-1}b + Tx^n =: \Phi(x^n).$$

T is the **iteration operator**. This is the form used to analyze convergence and rates of convergence.

(3) **Regular Splitting Form:** This form is similar to the last one for FOR. Rewrite

$$A = M - N, \quad \text{so} \quad N = M - A.$$

Then $Ax = b$ can be written $Mx = b + Nx$. The iteration is then

$$Mx^{n+1} = b + Nx^n.$$

For FOR $M = \rho I$ and $N + \rho I - A$ so the regular splitting form becomes

$$\rho I x^{n+1} = b + (\rho I - A)x^n.$$

Remark 6.2 (Jacobi method). As an example of a regular splitting form, consider the case that $A = D - L - U$, where D is the diagonal part of A, L is the lower triangular part of A, and U is the upper triangular part of A. The iteration in this case takes the form

$$Dx^{n+1} = b + (L + U)x^n.$$

This is called the Jacobi method.

6.1.2 *Three quantities of interest*

Detelina's Law:
"If your program doesn't run, that means it has an error in it.
If your program does run, that means it has two errors in it."

There are three important quantities to track in the iterative methods:

(1) The **error**, $e^n = x - x^n$, which is unknown but essential;
(2) The **residual**, $r^n = b - Ax^n$, which is computable; and,
(3) The **update**, $\Delta^n = x^{n+1} - x^n$, which is computable.

The residual and updates are used to give indications of the error and to decide when to stop an iterative method.

Theorem 6.1. *For first order Richardson* $(T = T_{FOR} = I - \rho^{-1}A)$, e^n, r^n *and* Δ^n *all satisfy the same iteration*

$$e^{n+1} = Te^n, \qquad r^{n+1} = Tr^n, \ and \quad \Delta^{n+1} = T\Delta^n.$$

Proof. This is by subtraction. Since

$$x = \rho^{-1}b + Tx \quad and$$
$$x^{n+1} = \rho^{-1}b + Tx^n,$$

subtraction gives

$$(x - x^{n+1}) = T(x - x^n) \qquad and \qquad e^{n+1} = Te^n.$$

For the update iteration, note that

$$x^{n+1} = \rho^{-1}b + Tx^n \quad and \qquad x^n = \rho^{-1}b + Tx^{n-1}.$$

Subtraction gives

$$(x^{n+1} - x^n) = T(x^n - x^{n-1}) \qquad and \qquad \Delta^{n+1} = T\Delta^n.$$

The residual update is a little trickier to derive. Since

$$\rho x^{n+1} = \rho x^n + b - Ax^n = \rho x^n + r^n$$

multiply by $-A$ and add ρb:

$$\rho\left(b - Ax^{n+1}\right) = \rho\left(b - Ax^n\right) - Ar^n$$
$$\rho r^{n+1} = \rho r^n - Ar^n \ and \ thus$$
$$r^{n+1} = (I - \rho^{-1}A)r^n = Tr^n.$$

$\square$

Remark 6.3. For other iterations, the typical result is

$$e^{n+1} = Te^n, \qquad \Delta^{n+1} = T\Delta^n, \ and \ r^{n+1} = ATA^{-1}r^n.$$

The matrix ATA^{-1} and T are similar matrices. For FOR, because of the special form of $T = \rho I - \rho^{-1}A$, $ATA^{-1} = T$ since

$$ATA^{-1} = A(I - \rho^{-1}A)A^{-1} = AA^{-1} - \rho^{-1}AAA^{-1} = I - \rho^{-1}A = T.$$

Thus, $r^{n+1} = Tr^n$.

This theorem has an important interpretation:

$$\Delta^n, r^n \quad \text{and} \quad e^n \longrightarrow 0 \quad \text{at the same rate.}$$

It is entirely possible that residuals r^n and errors e^n can be of widely different sizes. However, *since they both go to zero at the same rate, if the residuals improve by k significant digits from the initial residual, the errors will have typically also improved by k significant digits over the initial error.*

The third big question is **When to stop?** (Alternately, how to measure *"satisfaction"* with a computed answer.) The theorem is important because it says that monitoring the (computable) residuals and updates is a valid way to test it the (incomputable) error has been improved enough to stop.

Stopping Criteria: Every iteration should include three (!) tests of stopping criteria.

(1) **Too Many Iterations:** If $n \geq$ itmax (a user supplied maximum number of iterations), then the method is likely not converging to the true solution: stop, and signal failure.

(2) **Small Residual:** With a preset tolerance tol1 (e.g., tol1$= 10^{-6}$), test if:

$$\frac{\|r^n\|}{\|b\|} \leq \text{tol1}.$$

(3) **Small Update:** With tol2 a preset tolerance, test if:

$$\frac{\|\Delta^n\|}{\|x^n\|} \leq \text{tol2}.$$

The program should terminate if either the first test or *both* the second and third tests are satisfied. Usually other computable heuristics are also monitored to check for convergence and speed of convergence. One example is the experimental contraction constant

$$\alpha_n := \frac{\|r^{n+1}\|}{\|r^n\|} \quad \text{or} \quad \frac{\|\Delta^{n+1}\|}{\|\Delta^n\|}.$$

This is monitored because $\alpha_n > 1$ suggests divergence and $\alpha_n < 1$ but very close to 1 suggests very slow convergence.

To summarize, the important points

- Iterative methods require minimal storage requirements. They are essential for $3D$ problems!

- Basic iterative methods are easy to program.[3] The programs are short and easy to debug and often are inherently parallel.
- Iterative method's convergence can be fast or not at all. The questions of convergence (at all) and speed of convergence are essential ones that determine if an iterative method is practical or not.

6.2 Mathematical Tools

"The Red Queen: 'Why, sometimes I've believed as many as six impossible things before breakfast...'
Alice: 'Perhaps, but surely not all at the same time!'
The Red Queen: 'Of course all at the same time, or where's the fun?'
Alice: 'But that's impossible!'
The Red Queen: 'There, now that's seven!' "
— Lewis Carroll, Alice in Wonderland

To analyze the critically important problem of convergence of iterative methods, we will need to develop some mathematical preliminaries. We consider an $N \times N$ iteration matrix T and a fixed point x and the iteration x^n

$$x = b + Tx, \qquad x^{n+1} = b + Tx^n.$$

Subtracting we get the error equation

$$e^n = x - x^n \qquad satisfies \qquad e^{n+1} = Te^n.$$

Theorem 6.2. *Given the $N \times N$ matrix T, a necessary and sufficient condition that for any initial guess*

$$e^n \longrightarrow 0 \qquad as \quad n \to \infty$$

is that there exists a matrix norm $\| \cdot \|$ with

$$\|T\| < 1.$$

Proof. Sufficiency is easy. Indeed, if $\|T\| < 1$ we have

$$\|e^n\| = \|Te^{n-1}\| \le \|T\|\|e^{n-1}\|.$$

---

[3]Of course, in MATLAB direct solution is an intrinsic operator ($\backslash$), so even the simplest iterative methods are more complicated. Further, often the requirements of rapid convergence adds layers of complexity to what started out as a simple implementation of a basic iterative method.

Since $e^{n-1} = Te^{n-2}$, $\|e^{n-1}\| \leq \|T\|\|e^{n-2}\|$ so $\|e^n\| \leq \|T\|^2\|e^{n-2}\|$. Continuing backward (for strict proof, this means : using an induction argument) we find

$$\|e^n\| \leq \|T\|^n\|e^0\|.$$

Since $\|T\| < 1$, $\|T\|^n \to 0$ as $n \to \infty$.

Proving that convergence implies existence of the required norm is harder and will follow from the next two theorems that complete the circle of ideas.   □

The proof that $\|T\| < 1$ for some norm $\|\cdot\|$ is also mathematically interesting and important. It is implied by the next theorem.

Theorem 6.3. *For any $N \times N$ matrix T, a matrix norm $\|\cdot\|$ exists for which $\|T\| < 1$ if and only if for all eigenvalues $\lambda(T)$*

$$|\lambda(T)| < 1.$$

Definition 6.3 (spectral radius). The **spectral radius** of an $N \times N$ matrix T, $spr(T)$, is the size of the largest eigenvalue of T

$$spr(T) = \max\{|\lambda| : \lambda = \lambda(T)\}.$$

Theorem 6.4. *Given any $N \times N$ matrix T and any $\varepsilon > 0$ there exists a matrix norm $\|\cdot\|$ with $\|T\| \leq spr(T) + \varepsilon$.*

Proof. See Appendix A   □

Using this result, the following fundamental convergence theorem holds.

Theorem 6.5. *A necessary and sufficient condition that*

$$e^n \to 0 \quad as \quad n \to \infty \quad for \; any \quad e^0,$$

is that $spr(T) < 1$.

Proof. That it suffices follow from the previous two theorems. It is easy to prove that it is necessary. Indeed, suppose $\rho(T) \geq 1$ so T has an eigenvalue λ

$$T\phi = \lambda\phi \quad with \quad |\lambda| \geq 1.$$

Pick $e^0 = \phi$. Then, $e^1 = Te^0 = T\phi = \lambda\phi$, $e^2 = Te^1 = \lambda^2\phi, \ldots,$

$$e^n = \lambda^n\phi.$$

Since $|\lambda| \geq 1$, e^n clearly does not approach zero as $n \to \infty$.   □

Since the eigenvalues of T determine if the iteration converges, it is useful to know more about eigenvalues.

Definition 6.4 (similar matrices). B and PBP^{-1} are said to be similar matrices.

Lemma 6.1 (Similar matrices have the same eigenvalues). *Let B be any $N \times N$ matrix and P any invertible matrix then the similar matrices*

$$B \quad and \quad PBP^{-1}$$

have the same eigenvalues.

Proof. The proof is based on interpreting a similarity transformation as a change of variable:

$$B\phi = \lambda\phi \text{ holds if and only if}$$
$$PB(P^{-1}P)\phi = \lambda P\phi, \text{ if and only if}$$
$$(PBP^{-1})\Psi = \lambda\Psi, \text{ where } \Psi = P\phi.$$

$\square$

For many functions $f(x)$ we can insert an $N \times N$ matrix A for x and $f(A)$ will still be well defined as an $N \times N$ matrix. Examples include

$$f(x) = \frac{1}{x} \Rightarrow f(A) = A^{-1}$$

$$f(x) = \frac{1}{1+x} \Rightarrow f(A) = I + A^{-1}$$

$$f(x) = e^x = 1 + x + \frac{x^2}{2!} + \ldots \Rightarrow f(A) = e^A = \sum_{n=0}^{\infty} \frac{A^n}{n!},$$

$$f(x) = x^2 - 1 \Rightarrow f(A) = A^2 - I.$$

In general, $f(A)$ is well defined (by its power series as in e^A) for any analytic function. The next theorem, known as the *Spectral Mapping Theorem*, is extremely useful. It says that

the eigenvalues of the matrix $f(A)$ are f(the eigenvalues of A):

$$\lambda(f(A)) = f(\lambda(A)).$$

Theorem 6.6 (Spectral Mapping Theorem). *Let* $f : \mathbb{C} \to \mathbb{C}$ *be an analytic function.*[4] *If* (λ, ϕ) *is an eigenvalue, eigenvector pair for* A *then* $(f(\lambda), \phi)$ *is an eigenvalue, eigenvector pair for* $f(A)$.

Exercise 6.5. *(a) Let* A *be a* 2×2 *matrix with eigenvalues* 2, -3. *Find the eigenvalues of* e^A, A^3, $(2A+2I)^{-1}$. *For what values of* a, b *is the matrix* $B = aI + bA$ *invertible? Explain. (b) If the eigenvalues of a symmetric matrix* A *satisfy* $1 \le \lambda(A) \le 200$, *find an interval (depending on* ρ) *that contains the eigenvalues of* $\lambda(T)$, $T = I - (1/\rho)A$. *For what values of* ρ *are* $|\lambda(T)| < 1$? *(c) For the same matrix, find an interval containing the eigenvalues of* $(I + A)^{-1}(I - A)$.

Exercise 6.6. *If* A *is symmetric, show that* $cond_2(A^t A) = (cond_2(A))^2$.

Exercise 6.7. *Let* A *be invertible and* $f(z) = 1/z$. *Give a direct proof of the SMT for this particular* $f(\cdot)$. *Repeat for* $f(z) = z^2$.

6.3   Convergence of FOR

> I was a very early believer in the idea of convergence.
> — Jean-Marie Messier

This section gives a complete and detailed proof that the First Order Richardson iteration

$$\rho(x^{n+1} - x^n) = b - Ax^n \tag{6.1}$$

converges, for any initial guess, to the solution of

$$Ax = b.$$

The convergence is based on two essential assumptions: that the matrix A is symmetric, positive definite (SPD) and the parameter ρ is chosen large enough. The convergence proof will also give an important information on

- how large is "large enough",
- the optimal choice of ρ,
- the expected number of steps of FOR required.

---

[4]Analytic is easily weakened to analytic in a domain, an open connected set, including the spectrum of A.

Theorem 6.7 (Convergence of FOR). *Suppose A is SPD. Then FOR converges for any initial guess x_0 provided*

$$\rho > \lambda_{\max}(A)/2.$$

Proof of Theorem 6.7. Rewrite $Ax = b$ as $\rho(x - x) = b - Ax$. Subtracting (6.1) from this gives the error equation

$$\rho(e^{n+1} - e^n) = -Ae^n, \quad e^n = x - x^n,$$

or

$$e^{n+1} = Te^n, \quad T = (I - \rho^{-1}A).$$

From Section 6.2, we know that $e^n \to 0$ for any e^0 if and only if $|\lambda(T)| < 1$ for every eigenvalue λ of the matrix T. If $f(x) = 1 - x/\rho$, note that $T = f(A)$. Thus, by the spectral mapping theorem

$$\lambda(T) = 1 - \lambda(A)/\rho.$$

Since A is SPD, its eigenvalues are real and positive:

$$0 < a = \lambda_{\min}(A) \le \lambda(A) \le \lambda_{\max}(A) = b < \infty.$$

We know $e^n \to 0$ provided $|\lambda(T)| < 1$, or

$$-1 < 1 - \lambda(A)/\rho < +1.$$

Since $\lambda(A) \in [a, b]$, this is implied by

$$-1 < 1 - \frac{x}{\rho} < +1, \quad for \quad a \le x \le b.$$

This is true if and only if (for $\rho > 0$)

$$-\rho < \rho - x < +\rho$$

or $-2\rho < -x < 0$ or $0 < x < 2\rho$ or

$$\rho > \frac{x}{2} \quad for \ \ 0 < a \le x \le b.$$

This is clearly equivalent to

$$\rho > \frac{b}{2} = \frac{\lambda_{\max}(A)}{2}.$$

$\square$

6.3.1 *Optimization of ρ*

> If you optimize everything, you will always be unhappy.
> — Donald Knuth

Clearly, the smaller $\|T\|_2$ the faster $e^n \to 0$. Now, from the above proof

$$\|T\|_2 = \max |\lambda(T)| = \max |1 - \lambda(A)/\rho|.$$

The eigenvalues $\lambda(A)$ are a discrete set on $[a, b]$. A simple sketch (see the next subsection) shows that

$$\|T\|_2 = \max\{|1 - \lambda(A)/\rho| : \text{all } \lambda(A)\} =$$

$$\max\{|1 - \lambda_{\min}/\rho|, |1 - \lambda_{\max}/\rho|\}.$$

We see that $\|T\|_2 < 1$ for $\rho > b/2$, as proven earlier. Secondly, we easily calculate that the "optimal" value of ρ is

$$\alpha = \alpha_{\min} \quad \text{at} \quad \rho = \frac{a + b}{2} = \frac{\lambda_{\min} + \lambda_{\max}}{2}.$$

Let

$$\kappa = \frac{\lambda_{\max}}{\lambda_{\min}}$$

denote the spectral condition number of A. Then, if we pick

- $\rho = \lambda_{\max}(A)$, $\|T\|_2 = 1 - \frac{1}{\kappa}$,
- $\rho = (\lambda_{\max} + \lambda_{\min})/2$, $\|T\|_2 = 1 - \frac{2}{\kappa+1}$.

Getting an estimate of $\lambda_{\max}(A)$ is easy; we could take, for example,

$$\lambda_{\max} \le \|A\|_{\text{ANYNORM}} = \text{e.g.} = \max_{1 \le i \le N} \sum_{j=1}^{N} |a_{ij}|.$$

However, estimating $\lambda_{\min}(A)$ is often difficult. The shape of $\alpha(\rho)$ also suggests that it is better to overestimate ρ than underestimate ρ. Thus, often one simply takes $\rho = \|A\|$ rather than the "optimal" value of ρ. The cost of this choice is that it roughly doubles the number of steps required.

6.3.2 *Geometric analysis of the min-max problem*

The problem of selecting an optimal parameter for SPD matrices A is a one parameter min-max problem. There is an effective and insightful way to solve all such (one parameter min-max) problems by drawing a figure and saying "Behold!".[5] In this sub-section we shall solve the optimal parameter problem by this geometric approach. We shall give the steps in detail (possibly excruciating detail even) with apologies to the many readers for whom the curve sketching problem is an easy one.

Following the previous sections, we have that the error satisfies

$$e^{n+1} = T_\rho e^n = (I - \frac{1}{\rho}A)e^n.$$

FOR converges provided

$$|\lambda|_{\max}(T) = \max\{|1 - \lambda(A)/\rho| : \lambda \text{ an eigenvalue of A}\} < 1.$$

The parameter optimization problem is then to find $\rho_{optimal}$ by

$$\min_\rho \max_{\lambda=\lambda(A)} |1 - \lambda/\rho|.$$

To simplify this we suppose that only the largest and smallest eigenvalues (or estimates thereof) are known. Thus, let

$$0 < a = \lambda_{\min}(A) \leq \lambda \leq b = \lambda_{\max}(A) < \infty$$

so that the simplified parameter optimization problem is

$$\min_\rho \max_{a\leq\lambda\leq b} |1 - \lambda/\rho|.$$

Fix one eigenvalue λ and consider in the $y - \rho$ plane the curve $y = 1 - \lambda/\rho$, as in Figure 6.1. The plot also has small boxes on the ρ axis indicating $a = \lambda_{\min}(A), b = \lambda_{\max}(A)$ and the chosen intermediate value of λ.

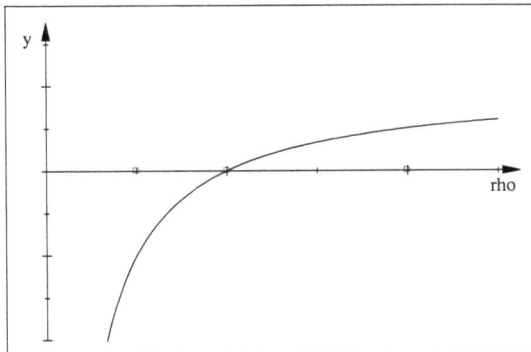

Fig. 6.1   Plot of $y = 1 - \lambda/\rho$ for one value of λ.

---

[5] All the better if one can say it in Greek.

The next step is for this same one eigenvalue λ to consider in the $y - \rho$ plane the curve $y = |1 - \lambda/\rho|$, as shown in Figure 6.2. This just reflects up the portion of the curve below the rho axis in the previous figure. We also begin including the key level $y = 1$.

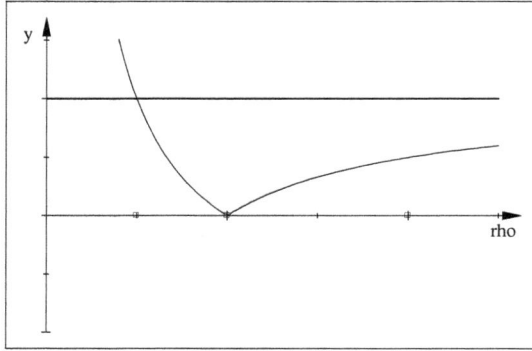

Fig. 6.2   Plot of $y = |1 - \lambda/\rho|$ for one value of λ.

The next step is to plot $y = \max_{a \leq \lambda \leq b} |1 - \lambda/\rho|$. This means plotting the same curves for a few more values of λ and taking the upper envelope of the family of curves once the pattern is clear. We do this in two steps. First Figure 6.3 includes more examples of $y = |1 - \lambda/\rho|$.

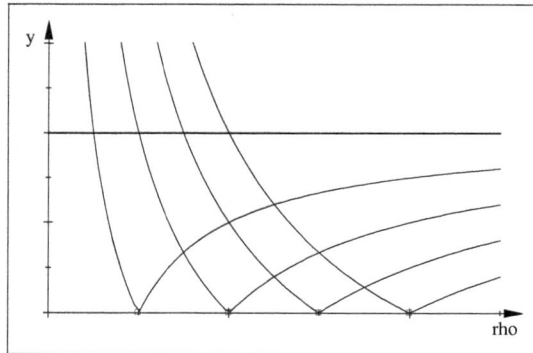

Fig. 6.3   The family $y = |1 - \lambda/\rho|$ for four values of λ.

The upper envelope is just whichever curve is on top of the family. We plot it in Figure 6.4 with the two curves that comprise it.

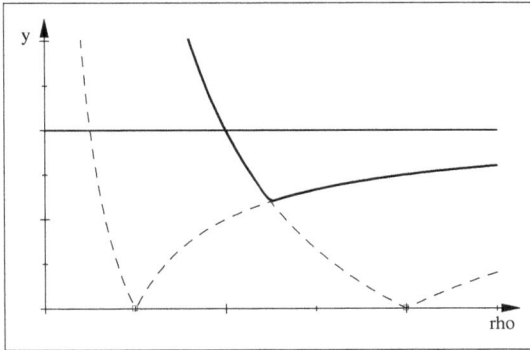

Fig. 6.4   The dark curve is $y = \max_{a \leq \lambda \leq b} |1 - \lambda/\rho|$.

The dark curve in Figure 6.4 is our target. It is

$$\|T(\rho)\|_2 = \max_{a \leq \lambda \leq b} |1 - \lambda/\rho|.$$

Checking which individual curve is the active one in the maximum, we find:

- **Convergence:** $\|T(\rho)\|_2 < 1$ if and only if ρ is bigger than the value of rho at the point where $1 = -(1 - \lambda_{\max}/\rho)$. Solving this equation for rho we find the condition

  convergence if and only if $\rho > \lambda_{\max}/2$.

- **Parameter selection:** The optimal value of rho is where $\min_\rho \max_{a \leq \lambda \leq b} |1 - \lambda/\rho|$ is attained. This is the value of rho where the dark, upper envelope curve is smallest. Checking the active constraints, it is where the two dashed curves cross and thus where $(1 - \lambda_{\min}/\rho) = -(1 - \lambda_{\max}/\rho)$. Solving for rho gives the value

  $$\rho_{optimal} = \frac{\lambda_{\min} + \lambda_{\max}}{2},$$

  the y value is given by

  $$\|T(\rho_{opotimal})\|_2 = 1 - \frac{2}{\kappa + 1},$$

  and the condition number is

  $$\kappa = cond_2(A) = \frac{\lambda_{\max}(A)}{\lambda_{\min}(A)}.$$

6.3.3 *How many FOR iterations?*

"Computing is no more about computers than astronomy is about telescopes."
— Edsger Dijkstra

The above analysis also gives insight on the expected number of iterations for FOR to converge. Since

$$e^n = Te^{n-1}, \text{ so we have } e^n = Te^0.$$

Because of the multiplicative property of norms

$$\|e^n\| \le \|T^n\|\|e^0\| \le \|T\|^n\|e^0\|.$$

Thus, the relative improvement in the error is

$$\frac{\|e^n\|}{\|e^0\|} \le \|T\|^n.$$

If we want the initial error to be improved by some factor ε then we want

$$\frac{\|e^n\|}{\|e^0\|} < \varepsilon.$$

Since $\|e^n\|/\|e^0\| \le \|T\|^n$, it suffices that $\|T\|^n < \varepsilon$ or (taking logs and solving for n)

$$n \ge \frac{\ln(\frac{1}{\varepsilon})}{\ln\left(\frac{1}{\|T\|}\right)}.$$

Usually, we take $\varepsilon = 10^{-1}$ and speak of number of iterations for each significant digit of accuracy. This is

$$n \ge \ln(10)/\ln\left(\frac{1}{\|T\|}\right).$$

We can estimate how big this is using

$$\|T\| = 1 - \alpha, \text{ where } \alpha \text{ is small,}$$

and $\ln(1 - \alpha) \doteq -\alpha + O(\alpha^2)$ (by Taylor series). This gives

$$\frac{1}{\left(\frac{1}{\|T\|}\right)} \doteq \alpha^{-1}, \text{ where } \|T\| = 1 - \alpha.$$

Refer to the previous $\alpha = 1/\kappa(A)$ for $\rho = (\lambda_{\max}(A) + \lambda_{\min}(A))/2$.

Conclusions

(1) For $\rho = \lambda_{\max}(A)$, FOR requires approximately

$$n \doteq \ln(10) \cdot \kappa(A) \text{ iterations}$$

per significant digit of accuracy.
(2) For $\rho = (\lambda_{\max}(A) + \lambda_{\min}(A))/2$, FOR requires approximately

$$n \doteq \ln(10)\frac{1}{2}(\kappa(A) + 1) \text{ iterations}$$

per significant digit of accuracy.
(3) For the model Poisson problem, $\kappa(A) \doteq O(h^{-2})$, this gives

$$n = O(h^{-2}) \text{ iterations}$$

per significant digit of accuracy.
(4) The problem of FOR is that it is *too slow*. On, e.g., 100×100 meshes it requires tens of thousands of iterations for each significant digit sought. Thus, in the hunt for "better" iterative methods, it is clear "better" means "faster" which means *fewer iterations per significant digit* which means find an iteration for which its iteration operator T satisfies $spr(T)$ is smaller that of FOR!

Exercise 6.8. *Let A be $N \times N$ and SPD. Consider FOR for solving $Ax = b$. Define the A-norm by:*

$$|x|_A := \sqrt{x^t A x} = \sqrt{< Ax, x >} = \sqrt{< x, Ax >}.$$

Give a complete convergence analysis of the FOR error in the A norm (paralleling our analysis). What is the optimal ρ? In particular show that for the optimal value of ρ

$$\|x - x^n\|_A \leq \left(\frac{\kappa - 1}{\kappa + 1}\right) \|x - x^{n-1}\|_A.$$

What is the number of iterations per significant digit for the MPP? If you prefer, you can explore this computationally instead of theoretically [Choose one approach: analysis or computations, not both].

Exercise 6.9. *Consider error in FOR yet again. Suppose one chooses 2 values of ρ and alternates with $\rho_1, \rho_2, \rho_1, \rho_2$, etc. Relabel the steps as follows:*

$$\rho_1(x^{n+1/2} - x^n) = b - Ax^n$$
$$\rho_2(x^{n+1} - x^{n+1/2}) = b - Ax^{n+1/2}.$$

Eliminate the half step to write this as an stationary iterative method [i.e., relate x^{n+1} to x^n]. Analyze convergence for SPD A. Can this converge faster with two different values of ρ than with 2 steps of one value of ρ? If you prefer, you can explore this computationally instead of theoretically [Choose one approach: analysis or computations. It will be most exciting if you work with someone on this problem with one person doing the analysis and the other the numerical explorations].

Exercise 6.10. *Consider the 2D Model Poisson Problem on a uniform mesh with $h = 1/(N+1)$, boundary condition: $g(x,y) = 0$, right hand side: $f(x,y) = x + y$.*

*(1) Take $h = 1/3$ and **write down** the 4×4 linear system in matrix vector form.*
*(2) Given an $N \times N$ mesh, let $u(i,j)$ denote an $N \times N$ array of approximations at each (x_i, y_j). **Give pseudocode** for computing the residual $r(i,j)$ ($N \times N$ array) and its norm. c. Suppose the largest N for which the coefficient matrix can be stored in banded sparse form (to be solved by Gaussian elimination) is $N = 150$.*
*(3) **Estimate** the largest value of N the problem can be stored to be solved by First Order Richardson. Explain carefully!*

6.4 Better Iterative Methods

"Computer scientists want to study computing for its own sake; computational scientists want to build useful things."
— Greg Wilson

FOR has a huge savings in storage over Gaussian elimination but not in time to calculate the solution. There are many better iterative methods; we consider a few such algorithms in this section: the Gauss–Seidel method, over-relaxation and the SOR method. These are still used today — not as solvers but as preconditioners for the Conjugate Gradient method of Chapter 7.

6.4.1 *The Gauss–Seidel Method*

"Euler published 228 papers after he died, making the deceased Euler one of history's most prolific mathematicians."
— William Dunham

The Gauss–Seidel Method is easiest to understand for the $2D$ model problem by comparing it with the Jacobi method (which is FOR with $\rho = 4$). The Jacobi or $\rho = 4$ FOR method is:

Algorithm 6.5 (Jacobi Algorithm for the 2D MPP). *Given an array* uold *of size* N+1 *by* N+1 *with boundary values filled with zeros, a maximum number of iterations* itmax, *and a tolerance* tol,

```
  h=1/N
  for it=1:itmax
    for i=2:N
      for j=2:N
(*)     unew(i,j)=h^2*f(i,j)+ ...
             ( uold(i+1,j)+uold(i-1,j)+ ...
               uold(i,j+1)+uold(i,j-1) )/4
      end
    end
    if convergence is satisfied, exit
    unew=uold
  end
```

The idea of Gauss–Seidel is to use the best available information instead: if unew is known at a neighbor in step (*), why not use it instead of uold? This even makes it simpler to program and reduces the storage needed because we no longer have to track old and new values and simply use the most recent one.

Algorithm 6.6 (Gauss–Seidel algorithm for the 2D MPP). *Given an array* u *of size* N+1 *by* N+1 *with boundary values filled with zeros, a maximum number of iterations* itmax, *and a tolerance* tol,

```
  h=1/N
  for it=1:itmax
    for i=2:N
      for j=2:N
(*)     u(i,j)=h^2*f(i,j)+ ...
             ( u(i+1,j)+u(i-1,j)+u(i,j+1)+u(i,j-1) )/4
      end
    end
    if convergence is satisfied, exit
  end
```

Algebraically, Jacobi and Gauss–Seidel for a general linear system are equivalent to splitting A into

$$A = D + L + U,$$
$$D = \text{diagonal of } A,$$
$$L = \text{lower triangular part of } A$$
$$U = \text{upper triangular part of } A.$$

$Ax = b$ is rewritten as $(L + D + U)x = b$. The Jacobi iteration for $Ax = b$ is

$$D\left(x^{n+1} - x^n\right) = b - Ax^n$$

equivalently: (Jacobi for Ax=b)

$$Dx^{n+1} = b - (L + U)x^n.$$

The Gauss–Seidel iteration for $Ax = b$ is

$$(D + U)\left(x^{n+1} - x^n\right) = b - Ax^n$$

equivalently: (Gauss–Seidel for Ax=b)

$$(D + U)x^{n+1} = b - Lx^n.$$

Both take the general form

pick M then:
$$M\left(x^{n+1} - x^n\right) = b - Ax^n.$$

There is a general theory for stationary iterative methods in this general form. The heuristics that are derived from this theory are easy to summarize:

- M must be chosen so that $spr(I - M^{-1}A) < 1$ and so that $M\Delta x = r$ is easy to solve.
- The closer M is to A the faster the iteration converges.

Costs for the MPP: In most cases, the Gauss–Seidel iteration takes approximately $\frac{1}{2}$ as many steps as Jacobi iteration. This is because, intuitively speaking, each time (*) in Algorithm 6.6 is executed, it involves half old values and half updated values. Thus, using Gauss–Seidel over FOR cuts execution time roughly in half. However, the model problem still needs $\frac{1}{2}O(h^{-2})$ iterations. Cutting costs by 50% is always good. However, the essential problem is how the costs grow as $h \to 0$. In other words, the goal should be to *cut the exponent as well as the constant!*

6.4.2 *Relaxation*

"Taniyama was not a very careful person as a mathematician. He made a lot of mistakes, but he made mistakes in a good direction, and so eventually, he got right answers. I tried to imitate him, but I found out that it is very difficult to make good mistakes."
— Goro Shimura, of the Shimura–Taniyama Conjecture

"The time to relax is when you don't have time for it."
— Sydney J. Harris

Relaxation is an ingenious idea. It is appealing because

• it is algorithmically easy to put into any iterative method program.
• it introduces a parameter that must be chosen. With the right choice, often it can reduce the number of iterations required significantly.

The second point (cost reduction) happens in cases where the number of steps is sensitive to the precise choice of the parameter. However, it is not appealing because

• it introduces a parameter which must be chosen problem by problem and the number of steps can increase dramatically for slightly non-optimal choices.

The idea is simple: pick the relaxation parameter ω, then add one line to an existing iterative solver as follows.

Algorithm 6.7 (Relaxation Step). *Given $\omega > 0$, a maximum number of iterations* itmax *and x^0:*

```
for n=1:itmax
```
    Compute x_{temp}^{n+1} by some iterative method
    Compute $x^{n+1} = \omega x_{temp}^{n+1} + (1 - \omega)x^n$
    if x^{n+1} is acceptable, exit
```
end
```

Since the assignment operator "=" means "replace the value on the left with the value on the right", in a computer program there is sometimes no need to allocate extra storage for the temporary variable x_{temp}^{n+1}. Under-relaxation means $0 < \omega < 1$ and is a good choice if the underlying iteration undershoots and overshoots in an alternating manner. Small positive values of ω can slow convergence to an impractical level, but rarely cause divergence. Over-relaxation means $\omega > 1$ and is a good choice when the underlying iteration is progressing slowly in a single direction. The right choice of ω can drastically improve convergence, but can cause divergence if ω is too big. This is because under relaxation is just linear interpolation between the past two values while over-relaxation is linear extrapolation from the past two values. For matrices arising from the MPP and similar problems, a theory for finding the optimal value of ω during the course of the iteration is well-established, see Hageman and Young [Hageman and Young (1981)].

Exercise 6.11. *In this exercise, you will see a simple example of how over-relaxation or under-relaxation can accelerate convergence of a sequence.*

For a number r with $|r| < 1$, consider the sequence[6] $\{e^n = (r)^n\}_{n=0}^{\infty}$. This sequence satisfies the recursion

$$e^n = re^{n-1} \tag{6.2}$$

and converges to zero at a rate r. Equation (6.2) can be relaxed as

$$e^n = \omega re^{n-1} + (1-\omega)e^n = (1 + \omega(r-1)e^{n-1}. \tag{6.3}$$

(1) Assume that $0 < r < 1$ is real, so that the sequence $\{e^n\}$ is of one sign. Show that there is a value ω_0 so that if $1 < \omega < \omega_0$, then (6.3) converges more rapidly than (6.2).

(2) Assume that $-1 < r < 0$ is real, so that the sequence $\{e^n\}$ is of alternating sign. Show that there is a value ω_0 so that if $0 < \omega_0 < \omega < 1$, then (6.3) converges more rapidly than (6.2).

(3) Assume that r is real, find the value ω_0 and show that, in this very special case, the relaxed expression converges in a single iteration.

Exercise 6.12. *Show that FOR with relaxation does not improve convergence. It just corresponds to a different value of ρ in FOR.*

Exercise 6.13. *Consider Gauss–Seidel plus relaxation (which is the SOR method studied next). Eliminate the intermediate (temporary) variable and show that the iteration operator is*

$$T(\omega) = \left(\frac{1}{\omega}D + L\right)^{-1}\left(\frac{1-\omega}{\omega}D - U\right).$$

[6]The notation $(r)^n$ means the nth power of r as distinct from e^n, meaning the nth iterate.

6.4.3 *Gauss–Seidel with over-relaxation = Successive Over Relaxation*

"The researches of many commentators have already thrown much darkness on this subject, and it is probable that if they continue we shall soon know nothing at all about it."
— Mark Twain

SOR = Successive Over Relaxation is one of the most famous algorithms in numerical analysis. It is simply Gauss–Seidel plus over relaxation. For many years it was the method of choice for solving problems like the model Poisson problem and its theory is both lovely and complete. Unfortunately, it also includes a tuning parameter that must be chosen. For the MPP it is tabulated and for a class of problems including MPP, methods for closely approximating the optimal ω are well-known. For more complex problems finding the optimal ω, while theory assures us that it exists, presents practical difficulties.

Heuristics exist for choosing good guesses for optimal ω, sometimes equation by equation. In industrial settings, there is a long history of refined heuristics based on theoretical results. In one case at Westinghouse, Dr. L. A. Hageman[7] was asked to devise automated methods for finding optimal ω values for a large computer program used for design of nuclear reactors. This program typically ran for hours and would often fail in the middle of the night, prompting a telephone call to the designer who had submitted the problem. The designer could sometimes get the program running again by reducing the chosen value of ω, otherwise, a late-night trip to work was required. In addition to the inconvenience, these failures caused loss of computer time, a limited and valuable resource at the time. Dr. Hageman relieved the users of estimating ω and failures of the program were largely limited to modelling errors instead of solution errors. Methods for estimating ω can be found in Hageman and Young [Hageman and Young (1981)].

For $A = L + D + U$ SOR is as follows:

Algorithm 6.8 (SOR for Ax=b). *Given $\omega > 0$, a maximum number of iterations* itmax *and* x^0:

Compute $r^0 = b - Ax^0$
```
for n=1:itmax
```

---

[7]Personal communication.

Compute x_{temp}^{n+1} by one GS step:
$$(D+U)\left(x_{temp}^{n+1} - x^n\right) = b - Ax^n$$
Compute $x^{n+1} = \omega x_{temp}^{n+1} + (1-\omega)x^n$
Compute $r^{n+1} = b - Ax^{n+1}$
if x^{n+1} and r^{n+1} are acceptable, exit
end

For the 2D MPP the vector x is the array $u(i,j)$ and the action of D, U and A can be computed directly using the stencil. That $D + U$ is upper triangular means just use the most recent value for any $u(i,j)$. It thus simplifies as follows.

Algorithm 6.9 (SOR algorithm for the 2D MPP). *Given an array* u *of size* N+1 *by* N+1 *with boundary values filled with zeros, a maximum number of iterations* itmax, *a tolerance* tol, *and an estimate for the optimal omega* =omega *(see below):*

```
h=1/N
for it=1:itmax
  for i=2:N
    for j=2:N
      uold=u(i,j)
      u(i,j)=h^2*f(i,j) ...
             + (u(i+1,j)+u(i-1,j)+u(i,j+1)+u(i,j-1))/4
      u(i,j)=omega*u(i,j)+(1-omega)*uold
    end
  end
  if convergence is satisfied, exit
end
```

Convergence results for SOR are highly developed. For example, the following is known.

Theorem 6.8 (Convergence of SOR). *Let A be SPD and let $T_{Jacobi} = D^{-1}(L+U)$ be the iteration matrix for Jacobi (not SOR). If $spr(T_{Jacobi}) < 1$, then SOR converges for any ω with $0 < \omega < 2$ and there is an optimal choice of ω, known as $\omega_{optimal}$, given by*

$$\omega_{optimal} = \frac{2}{1 + \sqrt{1 - (spr(T_{Jacobi}))^2}}.$$

For $\omega = \omega_{optimal}$ and T_{SOR}, $T_{GaussSeidel}$ the iteration matrices for SOR and Gauss–Seidel respectively, we have

$$spr(T_{SOR}) = \omega_{optimal} - 1 < spr(T_{GaussSeidel}) \leq (spr(T_{Jacobi}))^2 < 1.$$

The dramatic reason SOR was the method of choice for $\omega = \omega_{optimal}$ is that it reduces the *exponent* in the complexity estimate for the MPP

$$from \quad O(h^{-2}) \quad to \quad O(h^{-1}).$$

Exercise 6.14. *Theorem 5.3 presents the eigenvalues of the 3D MPP matrix, and the analogous expression for the 2D MPP (A) is*

$$\lambda_{pq} = 4\left(\sin^2\left(\frac{p\pi}{2(N+1)}\right) + \sin^2\left(\frac{q\pi}{2(N+1)}\right)\right),$$
$$for \quad 1 \leq p, q \leq N.$$

Using this expression along with the observation that the diagonal of A is a multiple of the identity, find $spr(T_{Jacobi})$ and $spr(T_{SOR})$ for $\omega = \omega_{optimal}$. How many iterations will it take to reduce the error from 1 to 10^{-8} using: (a) Jacobi, and (b) SOR with $\omega = \omega_{optimal}$ for the case that $N = 1000$?

6.4.4 *Three level over-relaxed FOR*

"I guess I should warn you if I turn out to be particularly clear, you've probably misunderstood what I said."
— Alan Greenspan, at his 1988 confirmation hearings.

Adding a relaxation step to FOR just results in FOR with a changed value of ρ. It is interesting that if a relaxation step is added to a 2 stage version of FOR, it can dramatically decrease the number of steps required. The resulting algorithm is often called Second Order Richardson and it works like this:

Algorithm 6.10 (Second order Richardson). *Given the matrix A, initial vector u^0, values of ρ and ω, and a maximum number of steps* itmax:

Do one FOR step:
$$r^0 = b - Au^0$$
$$u^1 = u^0 + r^0/\rho$$
for n = 1:itmax
$$r^n = b - Au^n$$
$$u_{TEMP}^{n+1} = u^n + r^n/\rho$$

$$u^{n+1} = \omega u^{n+1}_{TEMP} + (1 - \omega)u^{n-1}$$

if *converged, exit,* end

end

It has some advantages to the other SOR on some parallel architectures (and some other disadvantages as well, such as having to optimize over two parameters).

It can reduce the number of iterations as much as SOR. It takes more to program it and requires more storage than SOR. However, it is parallel for the model problem while SOR is less so. In 2 stage algorithms, it is usual to name the variables u^{OLD}, u^{NOW} and u^{NEW}.

Algorithm 6.11 (Second order Richardson for the MPP). *Given a maximum number of iterations* itmax, *an* $(N + 1) \times (N + 1)$ *mesh on a square, starting with guesses* uold(i,j), unow(i,j) *and choices of* $\rho =$ rho, *and* $\omega =$ omega

```
for its=1:itmax
  for i=2:N
    for j=2:N
      au = - uold(i+1,j) - uold(i-1,j) ...
           + 4.0*uold(i,j) ...
           - uold(i,j+1) - uold(i,j-1)
      r(i,j) = h^2*f(i,j) - au
      unow(i,j) = uold(i,j) + (1/rho)*r(i,j)
      unew(i,j) = omega*unow(i,j) + (1-omega)*uold(i,j)
    end
  end
  Test for convergence
  if convergence not satisfied
    Copy unow to uold and unew to unow
    for i=2:N
      for j=2:N
        uold(i,j)=unow(i,j)
        unow(i,j)=unew(i,j)
      end
    end
  else
    Exit with converged result
```

```
  end

end
```

Convergence analysis has been performed for two stage methods.[8]

6.4.5   *Algorithmic issues: storing a large, sparse matrix*

"Just as there are wavelengths that people cannot see, and sounds that people cannot hear, computers may have thoughts that people cannot think."
— Richard Hamming, a pioneer numerical analyst

Neo: The Matrix.
Morpheus: Do you want to know what it is?
Neo: Yes.
Morpheus: The Matrix is everywhere.
— an irrelevant quote from the film "The Matrix"

If you are using an iterative method to solve $Ax = b$, most typically the method will be written in advance but all references to A will be made through a function or subroutine that performs the product Ax. It is in this subroutine that the storage for the matrix A is determined. The "best" storage scheme for very large systems is highly computer dependent. And, there are problems for which A need not be stored at all.

For example, for the 2D Model Poisson problem, a residual (and its norm) can be calculated on the physical mesh as follows.

Algorithm 6.12 (Calculating a residual: MPP). *Given an array* u *of size* N+1 *by* N+1 *containing the current values of the iterate, with correct values in boundary locations,*

```
rnorm=0
for i=2:N
  for j=2:N
    au=4*u(i,j)-(u(i+1,j)+u(i-1,j)+u(i,j+1)+u(i,j-1))
    r(i,j)=h^2*f(i,j)-au
```

---

[8]N.K. Nichols, On the convergence of two-stage iterative processes for solving linear equations, SIAM Journal on Numerical Analysis, 10 (1973), 460–469.

```
    rnorm=rnorm+r(i,j)^2
  end
end
rnorm=sqrt(rnorm/(N-1)^2)
```

Note that because the nonzero entries are known and regular the above did not even need to store the nonzero entries in A. We give one important example of a storage scheme for more irregular patterned matrices: **CRS=Compressed Row Storage.**

Example 6.2 (CRS=Compressed Row Storage). Consider a sparse storage scheme for the following matrix A below (where "$\cdots$" means *all the rest zeroes*).

$$
\begin{bmatrix}
2 & -1 & 0 & 0 & 3 & 0 & 0 & \cdots \\
0 & 2 & 0 & 1 & 0 & 0 & 5 & \cdots \\
-1 & 2 & -1 & 0 & 1 & 0 & 0 & \cdots \\
0 & 0 & 3 & 2 & 1 & 0 & 1 & \cdots \\
& & & \cdots & & &
\end{bmatrix}.
$$

To use A we need to first store the nonzero entries. In CRS this is done, row by row, in a long vector. If the matrix has M nonzero entries we store them in an array of length M named `value`

$$\text{value} = [2, -1, 3, 2, 1, 5, -1, 2, -1, 1, 3, 2, 1, 1, \ldots].$$

Next we need to know in the above vector the index where each row starts. For example, the first 3 entries, $2, -1, 3$, come from row 1 in A. Row 2 starts with the next (4th in this example) entry. This metadata can be stored in an array of length M named `row`, containing indices where each row starts. Of course, the first row always starts with the first value in `value`, so there is no need to store the first index, 1, leaving $(M-1)$ row indices to be stored. By convention, the final index in `value` is $(M+1)$.

$$\text{row} = [4, 7, 11, \ldots, M).$$

Now we know that Row 1 contains entries $1, 2, 3$ (because Row 2 starts with entry 4), we need to store the column numbers that each entry in `value` corresponds with in the global matrix A. This information can be stored in a vector of length M named `col`.

$$\text{col} = [1, 2, 5, 2, 4, 7, \ldots].$$

With these three arrays we can calculate the matrix vector product as follows.

Algorithm 6.13 (Matrix-Vector product with CRS). *Given the N-vector x and the $N \times N$ matrix A stored in CRS, this computes the N-vector $y = Ax$.*

```
first=1
for i=1:N
  y(i)=0
  for j=first:row(i)-1
    k=col(j)
    y(i)= y(i) + value(j)*x(k)
  end
  first=row(i)
end
```

Exercise 6.15. *Write a pseudocode routine for calculating $x \to A^t x$ when A is stored in CRS. Compare your routine to the one given in in the Templates book, Barrett, Berry, et al. [Barrett et al. (1994)], Section 4.3.2.*

Remark 6.4. The MATLAB program uses a variant of CRS to implement its sparse vector and matrix capabilities. See Gilbert, Moler and Schreiber [Gilbert *et al.* (1992)] for more information.

6.5 Dynamic Relaxation

There is one last approach related to stationary iterative methods we need to mention: Dynamic Relaxation. Dynamic relaxation is very commonly used in practical computing. In many cases for a given practical problem both the evolutionary problem $(x'(t) + Ax(t) = b)$ and the steady state problem $(Ax = b)$ must eventually be solved. In this case it saves programmer time to simply code up a time stepping method for the evolutionary problem and time step to steady state to get the solution of the steady problem $Ax = b$. It is also roughly the same for both linear and nonlinear systems, a highly valuable feature. Excluding programmer effort, however, it is however almost never competitive with standard iterative methods for solving linear systems $Ax = b$. To explain in the simplest case, suppose A is SPD and consider the linear system $Ax = b$. We embed this into the time dependent system of ODEs

$$\begin{aligned} x'(t) + Ax(t) &= b, \quad \text{for } 0 < t < \infty \\ x(0) &= x_0 \quad \text{the initial guess.} \end{aligned} \tag{IVP}$$

Since A is SPD it is not hard to show that as $t \to \infty$ the solution $x(t) \to A^{-1}b$.

Theorem 6.9. *Let A be SPD. Then for any initial guess the unique solution to (IVP) $x(t)$ converges to the unique solution of the linear system $Ax = b$:*

$$x(t) \to A^{-1}b \ \ as \ \ t \to \infty.$$

Thus one way to solve the linear system is to use any explicit method for the IVP and time step to steady state. There is in fact a $1-1$ correspondence between time stepping methods for some initial value problem associated with $Ax = b$ and stationary iterative methods for solving $Ax = b$. While this sounds like a deep meta-theorem it is not. Simply identify the iteration number n with a time step number and the correspondence emerges. For example, consider FOR

$$\rho(x^{n+1} - x^n) = b - Ax^n.$$

Rearrange FOR as follows:

$$\frac{x^{n+1} - x^n}{\Delta t} + Ax^n = b \text{ where } \Delta t := \rho^{-1}.$$

This shows that FOR is exactly the forward Euler method for IVP with timestep and pseudo-time

$$\Delta t := \rho^{-1} \ \text{ and } \ t^n = n\Delta t \ \text{ and } x^n \simeq x(t^n).$$

Similarly, the linear system $Ax = b$ can be embedded into a second order equation with damping

$$x''(t) + ax'(t) + Ax(t) = b, \quad \text{for } a > 0 \text{ and } 0 < t < \infty$$
$$x(0) = x_0, x'(0) = x_1 \quad \text{the initial guesses.}$$

Timestepping gives an iterative method with 2 parameters $(a, \Delta t)$ and thus resembles second order Richardson.

$$\frac{x^{n+1} - 2x^n + x^{n-1}}{\Delta t^2} + a\frac{x^{n+1} - x^{n-1}}{2\Delta t} + Ax^n = b .$$

The reasons this approach is not competitive, if programmer time is not counted, include:

- The evolutionary problem is forced to compute with physical time whereas an iterative method can choose some sort of pseudo-time that leads to steady state faster.

- The evolutionary problem seeks time accuracy to the preselected problem whereas iterative methods only seek to get to the steady state solution as fast as possible.

Exercise 6.16. *Find the IVP associated with the stationary iterative methods Gauss–Seidel and SOR.*

Exercise 6.17. *Complete the connection between second order Richardson and the second order IVP.*

Exercise 6.18. *Show that the solution of both IVP's converges to the solution of $Ax = b$ as $t \to \infty$.*

6.6 Splitting Methods

The classic and often very effective use of dynamic relaxation is in splitting methods. Splitting methods have a rich history; entire books have been written to develop aspects of them so we shall give one central and still important example, the Peaceman–Rachford method. Briefly, the $N \times N$ matrix A is split as

$$A = A_1 + A_2$$

where the subsystems $A_1 y = \mathrm{RHS}_1$ and $A_2 y = \mathrm{RHS}_2$ are "easy to solve". Usually easy to solve means easy either in computer time or in programmer effort; often A is split so that A_1 and A_2 are tridiagonal (or very close to tridiagonal) or so that you already have a code written to solve the subsystems that is highly adapted to their specific features. Given that the uncoupled problems can be solved, splitting methods then are applied to solve the coupled problems such as[9]

$$(A_1 + A_2)x = b,$$

$$\frac{d}{dt}x(t) + (A_1 + A_2)x(t) = f(t).$$

We consider the first two problems. We stress that splitting methods involve two separate steps and each is important for the success of the whole method:

- Pick the actual splitting $A = A_1 + A_2$.
- Pick the splitting method to be used with that splitting.

---

[9]Another possibility would be $F_1(x) + F_2(x) = 0$, where $A_i = F_i'$.

The first splitting method and in many ways still the best is the *Peaceman–Rachford* method.

Algorithm 6.14 (Peaceman–Rachford Method). *Pick parameter $\rho > 0$. Pick initial guess x^0.*

Until satisfied: given x^n

Solve

$$(\rho I + A_1)x^{n+1/2} = b + (\rho I - A_2)x^n, \qquad \text{(PR, step 1)}$$
$$(\rho I + A_2)x^{n+1} = b + (\rho I - A_1)x^{n+1/2}. \qquad \text{(PR, step 2)}$$

Test for convergence.

Each step is consistent with $Ax = b$. (If $x^{n+1/2} = x^n = x$ then rearranging $(\rho I + A_1)x^{n+1/2} = b + (\rho I - A_2)x^n$ gives $Ax = b$.) The first half step is A_1 implicit and A_2 explicit while the second half step reverses and is A_2 implicit and A_1 explicit. The classic questions are:

- When does it converge?
- How fast does it converge?
- How to pick the methods parameter?

We attack these by using the fundamental tools of numerical linear algebra.

Lemma 6.2. *The Peaceman–Rachford method satisfies*

$$x^{n+1} = \widetilde{b} + Tx^n$$

where the iteration operator $T = T_{PR}$ is

$$T_{PR} = T_{PR}(\rho) = (\rho I + A_2)^{-1}(\rho I - A_1)(\rho I + A_1)^{-1}(\rho I - A_2).$$

Proof. Eliminating the intermediate variable $x^{n+1/2}$ gives this immediately. $\qquad\square$

Thus Peaceman–Rachford converges if and only if[10] $spr(T_{PR}) < 1$. This is a product of four terms. It can be simplified to a product of the form $F(A_1) \cdot F(A_2)$ by commuting the (non-commutative) terms in the product using the following observation.

Lemma 6.3 (AB similar to BA). *Let A, B be $N \times N$ matrices. If either A or B is invertible, then AB is similar to BA*

$$AB \sim BA.$$

---

[10]Recall the spectral radius is $spr(T) := \max\{|\lambda|: \lambda$ is an eigenvalue of $T\}$.

Thus

$$spr(AB) = spr(BA).$$

Proof. Exercise! □

Define the function[11] $T : \mathbb{C} \to \mathbb{C}$ by

$$T(z) = \frac{\rho - z}{\rho + z}.$$

Using the property that we can commute matrices without altering the spectral radius of the product, we find

$$spr(T_{PR}(\rho)) = spr\left[(\rho I - A_1)(\rho I + A_1)^{-1}(\rho I - A_2)(\rho I + A_2)^{-1}\right]$$
$$= spr\left[T(A_1)T(A_2)\right].$$

We are now ready to prove one of the most famous results in iterative methods.

Theorem 6.10 (Kellogg's lemma). *Let B be an $N \times N$ real matrix. If*

$$x^T B x > 0 \text{ for all } 0 \neq x \in \mathbb{R}^N,$$

then

$$||T(B)||_2 < 1.$$

Proof. Let $x \neq 0$ be given. Then

$$\frac{||T(B)x||_2^2}{||x||_2^2} = \frac{\langle T(B)x, T(B)x \rangle}{\langle x, x \rangle}$$
$$= \frac{\langle (\rho I - B)(\rho I + B)^{-1}x, (\rho I - B)(\rho I + B)^{-1}x \rangle}{\langle x, x \rangle}.$$

Now change variables by $y = (\rho I + B)^{-1}x$, so $x = (\rho I + B)y$. We then have

$$\frac{||T(B)x||_2^2}{||x||_2^2} = \frac{\langle (\rho I - B)y, (\rho I - B)y \rangle}{\langle (\rho I + B)y, (\rho I + B)y \rangle}$$
$$= \frac{\rho^2 ||y||_2^2 - 2\rho y^T B y + ||By||_2^2}{\rho^2 ||y||_2^2 + 2\rho y^T B y + ||By||_2^2}.$$

Checking the numerator against the denominator and recalling $x^T B x > 0$, they agree term by term with one minus sign on top and the corresponding sign a plus on bottom. Thus

$$\frac{||T(B)x||_2^2}{||x||_2^2} < 1 \text{ and } ||T(B)||_2 \leq 1.$$

---

[11]This is an abuse of notation to use T for so many things. However, it is not too confusing and standard in the area (so just get used to it).

To prove strict inequality, assume equality holds. Then if $||T(B)||_2 = 1$, there must exist at least one $x \neq 0$ for which $||T(B)x||_2^2 = ||x||_2^2$. The same argument shows that for this x, $x^T B x = 0$, a contradiction. $\quad\square$

Using Kellogg's lemma we thus have a very strong convergence result for (PR).

Theorem 6.11 (Convergence of Peaceman–Rachford). *Let $\rho > 0$, $x^T A_i x \geq 0$ for $i = 1, 2$ with $x^T A_i x > 0$ for all $0 \neq x \in \mathbb{R}^N$ for one of $i = 1$ or 2. Then (PR) converges.*

Proof. We have

$$spr(T_{PR}) = spr\left[T(A_1)T(A_2)\right].$$

By Kellogg's lemma we have for both $i = 1, 2$, $||T(A_i)||_2 \leq 1$ with one of $||T(A_i)||_2 < 1$. Thus, $||T(A_1)T(A_2)||_2 < 1$ and

$$spr(T(A_1)T(A_2)) \leq ||T(A_1)T(A_2)||_2 < 1.$$

$\quad\square$

Remark 6.5 (What does the condition $x^T A_i x > 0$ mean?).
Peaceman–Rachford is remarkable in that its convergence is completely insensitive to the skew symmetric part of A. To see this, recall that any matrix can be decomposed into the sum of its symmetric part and its skew-symmetric part $A = A^s + A^{ss}$ by

$$A^s = \frac{1}{2}(A + A^T), \quad \text{so} \quad (A^s)^T = A^s$$

$$A^{ss} = \frac{1}{2}(A - A^T), \quad \text{so} \quad (A^{ss})^T = -A^{ss}.$$

For x a real vector it is not hard to check[12] that $x^T A^{ss} x \equiv 0$. Thus, $x^T A_i x > 0$ means:

The symmetric part of A is positive definite. The skew symmetric part of A is arbitrary and can thus be arbitrarily large.

Exercise 6.19. *The following iteration is silly in that each step costs as much as just solving $Ax = b$. Nevertheless, (and ignoring this aspect of it) prove convergence for matrices A with $x^T A x > 0$ and analyze the optimal parameter:*

$$w(\rho I + A)x^{n+1} = b + \rho x^n.$$

---

[12] $a := x^T A^{ss} x = \left(x^T A^{ss} x\right)^T = x^T A^{ssT} x \equiv -x^T A^{ss} x$. Thus $a = -a$ so $a = 0$.

Exercise 6.20. *Analyze convergence of the Douglas–Rachford method given by:*

$$(\rho I + A_1)x^{n+1/2} = b + (\rho I - A_2)x^n,$$
$$(\rho I + A_2)x^{n+1} = A_2 x^n + \rho x^{n+1/2}.$$

6.6.1   Parameter selection

The *Peaceman–Rachford* method requires a one parameter optimization problem be solved to pick ρ. We shall use exactly the same method as for FOR to solve the problem for SPD A. The solution process is exactly the same as for FOR but its conclusion is quite different, as we shall see. In this section we shall thus assume further that

$$A, A_1, A_2 \text{ are SPD.}$$

We shall actually solve the following problem where B plays the role of A_1, A_2:

Problem 6.1. Given the $N \times N$ SPD matrix B, find

$$\rho_{optimal} = \arg \min_{\rho} ||T(B)||_2, \text{ or}$$

$$\rho_{optimal} = \arg \min_{\rho} \max_{\lambda_{\min}(B) \leq \lambda \leq \lambda_{\max}(B)} |\frac{\rho - \lambda}{\rho + \lambda}|.$$

Consider $\phi(\rho) = |\frac{\rho - \lambda}{\rho + \lambda}|$, we follow the same steps as for FOR and sketch the curves below for several values of λ. We do this in two steps. In Figure 6.5 we plot more examples of $y = |\frac{\rho - \lambda}{\rho + \lambda}|$.

The upper envelope is the curve is on top of the family. We plot it in Figure 6.6 with the two curves that comprise it.

Solving for the optimal value by calculating the intersection point of the two curves comprising the upper envelop, we find

$$\rho_{optimal} = \sqrt{\lambda_{\max}\lambda_{\min}}$$
$$||T_{PR}(\rho_{optimal})||_2 = \frac{\sqrt{\kappa} - 1}{\sqrt{\kappa} + 1} = 1 - \frac{2}{\sqrt{\kappa} + 1}.$$

6.6.2   Connection to dynamic relaxation

The PR method can be written in residual-update form by eliminating the intermediate step and rearranging. There results

$$\frac{1}{2}(\rho I + A_1)(\rho I + A_2)\left(x^{n+1} - x^n\right) = b - Ax^n. \qquad \text{(PR, step 1)}$$

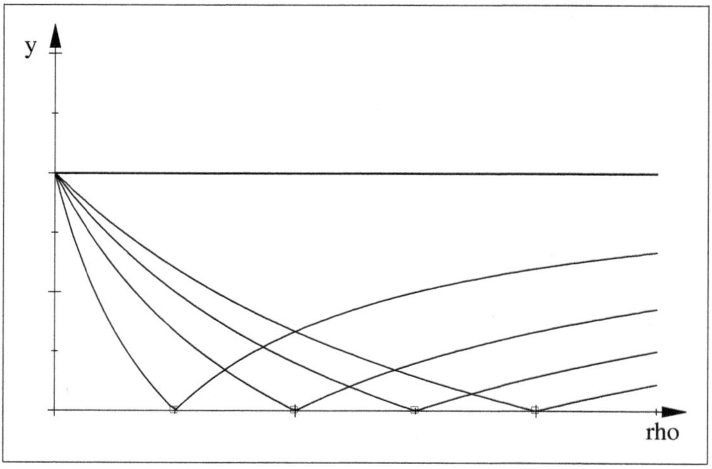

Fig. 6.5   The family $y = |\frac{\rho-\lambda}{\rho+\lambda}|$ for four values of λ.

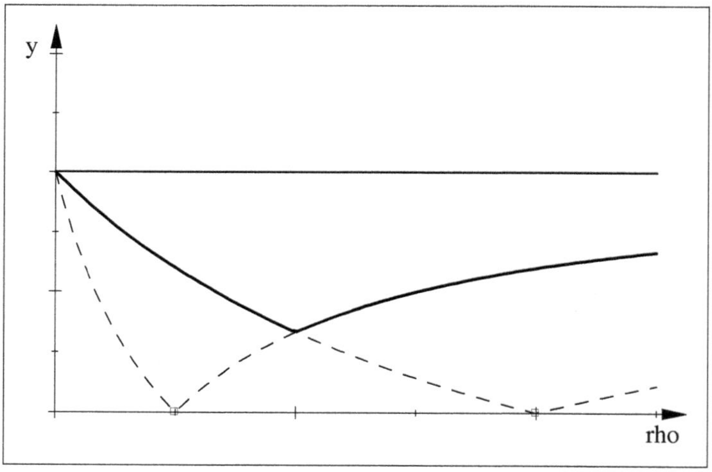

Fig. 6.6   The dark curve is $y = \max_{a \leq \lambda \leq b} |\frac{\rho-\lambda}{\rho+\lambda}|$.

6.6.3   *The ADI splitting*

To motivate the first (and still important) splitting $A = A_1 + A_2$, we recall a remark from the Gaussian elimination chapter.

Remark 6.6 (How fast is tridiagonal Gaussian elmination?).
Tridiagonal Gaussian elimination has 1 loop. Inside each loop roughly 3
arithmetic operations are performed. Thus, $O(N)$ floating point operations
are done inside tridiagonal Gaussian elimination for an $N \times N$ matrix. If
we solve an $N \times N$ linear system with a *diagonal* matrix, it will take N
divisions (one for each diagonal entry). The operation count of $3N - 3$
multiplies and divides for tridiagonal elimination is remarkable. Solving a
tridiagonal linear system is almost as fast as solving a diagonal (completely
uncoupled) linear system.

Indeed, consider the $2D$ MPP. Recall that the domain is the unit square,
$\Omega = (0,1) \times (0,1)$. Approximate u_{xx} and u_{yy} by

$$u_{xx}(a,b) \doteq \frac{u(a+\Delta x,b) - 2u(a,b) + u(a-\Delta x,b)}{\Delta x^2}, \quad (6.4)$$

$$u_{yy}(a,b) \doteq \frac{u(a,b+\Delta y) - 2u(a,b) + u(a,b-\Delta y)}{\Delta y^2}. \quad (6.5)$$

Introduce a uniform mesh on Ω with $N+1$ points in both directions: $\Delta x = \Delta y = \frac{1}{N+1} =: h$ and

$$x_i = ih, \quad y_j = jh, \quad i,j = 0,1,\ldots,N+1.$$

Let u_{ij} denote the approximation to $u(x_i, y_j)$ we will compute at each mesh
point. On the boundary use

$$u_{ij} = g(x_i, y_j) \quad (\text{ here } g \equiv 0) \quad \text{for each } x_i, y_j \text{ on } \partial\Omega$$

and eliminate the boundary points from the linear system. For a typical
(x_i, y_j) inside Ω we use

$$-\left(\frac{u_{i+1j} - 2u_{ij} + u_{i-1j}}{h^2} + \frac{u_{ij+1} - 2u_{ij} + u_{ij-1}}{h^2} \right) = f(x_i, y_j) \quad (6.6)$$

for all (x_i, y_j) inside of Ω

$$u_{ij} = g(x_i, y_j) \quad (\equiv 0) \quad \text{at all} \quad (x_i, y_j) \text{ on } \partial\Omega. \quad (6.7)$$

The boundary unknowns can be eliminated giving an $N^2 \times N^2$ linear system
for the N^2 unknowns:

$$A_{N^2 \times N^2} \, u_{N^2 \times 1} = f_{N^2 \times 1}.$$

To split A with the ADI = Alternating Direction Implicit splitting we use
the directional splitting already given above:

$$A = A_1 + A_2, \text{ where}$$

$$A_1 = -\frac{u_{i+1j} - 2u_{ij} + u_{i-1j}}{h^2}$$

$$A_2 = -\frac{u_{ij+1} - 2u_{ij} + u_{ij-1}}{h^2}.$$

Remark 6.7. *Solving* $(\rho I + A_i)v = RHS$ *requires solving one* $N \times N$, *tridiagonal linear system per horizontal mesh line (when* $i = 1$ *) or vertical mesh line (when* $i = 2$ *).* Solving tridiagonal linear systems is very efficient in both time and storage; one Peaceman–Rachford step with the ADI splitting is of comparable cost to 6 FOR steps.

Exercise 6.21. *If one full PR-ADI step costs the same as 6 FOR steps, is it worth doing PR-ADI? Answer this question using results on condition numbers of* $tridiag(-1, 2, -1)$ *and the estimates of number of steps per significant digit for each method.*

Chapter 7

Solving $Ax = b$ by Optimization

"According to my models, we are doubling the paradigm shift rate approximately every decade."
— From a letter to Scientific American by Ray Kurzweil

"Fundamentals, fundamentals. If you don't have them you'll run into someone else's."
— Virgil Hunter (Boxing trainer)

Powerful methods exist for solving $Ax = b$ when A is SPD based on a deep connection to an optimization problem. These methods are so powerful that often the best methods for solving a general linear system $Bx = f$ is to pass to the least squares equations $(B^t B) x = B^t f$ in which the coefficient matrix $A := B^t B$ is now coerced to be SPD (at the expense of squaring its condition number). We begin to develop them in this chapter, starting with some background.

Definition 7.1 (SPD matrices). $A_{N \times N}$ is **SPD** if it is **symmetric**, that is $A = A^t$, and **positive definite**, that is $x^t A x > 0$ for $x \neq 0$.

If A and B are symmetric we say $A > B$ if $A - B$ is SPD, i.e., if $x^t A x > x^t B x$ for all $x \neq 0$.

A is **negative definite** if $-A$ is positive definite.

A is **nonsymmetric** if $A \neq A^t$, **skew-symmetric** if $A^t = -A$ and **indefinite** if there are choices of x for which $x^t A x$ is both positive and negative.

A nonsymmetric (real) matrix A satisfying $x^t A x > 0$ for all real vectors $x \neq 0$ is called **positive real**.

It is known that a symmetric matrix A is positive definite if and only if all $\lambda(A) > 0$.

Lemma 7.1 (The A-inner product). *If A is SPD then*

$$\langle x, y \rangle_A = x^t A y$$

is a weighted inner product on $\mathbb{R}^n$, the A-inner product, and

$$\|x\|_A = \sqrt{\langle x, x \rangle_A} = \sqrt{x^t A x}$$

is a weighted norm, the A-norm.

Proof. $\langle x, y \rangle_A$ is bilinear:

$$\langle u + v, y \rangle_A = (u + v)^t A y = u^t A y + v^t A y = \langle u, y \rangle_A + \langle v, y \rangle_A$$

and

$$\langle \alpha u, y \rangle_A = (\alpha u)^t A y = \alpha \left(u^t A y \right) = \alpha \langle u, y \rangle_A.$$

$\langle x, y \rangle_A$ is positive

$$\langle x, x \rangle_A = x^t A x > 0, \text{ for } x \neq 0, \text{ since } A \text{ is SPD}.$$

$\langle x, y \rangle_A$ is symmetric:

$$\langle x, y \rangle_A = x^t A y = (x^t A y)^t = y^t A^t x^{tt} = y^t A x = \langle y, x \rangle_A.$$

Thus $\langle x, x \rangle_A$ is an inner product and, as a result, $\|x\|_A = \sqrt{\langle x, x \rangle_A}$ is an induced norm on $\mathbb{R}^N$. □

We consider two examples that show that the A-norm is a weighted ℓ_2 type norm.

Example 7.1. $A = \begin{bmatrix} 1 & 0 \\ 0 & 2 \end{bmatrix}$. Then the A norm of $[x_1, x_2]^t$ is

$$\|x\|_A = \sqrt{[x_1, x_2] \begin{bmatrix} 1 & 0 \\ 0 & 2 \end{bmatrix} \begin{bmatrix} x_1 \\ x_2 \end{bmatrix}} = \sqrt{x_1^2 + 2x_2^2},$$

which is exactly a weighted ℓ_2 norm.

Example 7.2. Let $A_{2 \times 2}$ be SPD with eigenvalues λ_1, λ_2 (both positive) and orthonormal eigenvectors ϕ_1, ϕ_2:

$$A \phi_j = \lambda_j \phi_j.$$

Let $x \in \mathbb{R}^2$ be expanded

$$x = \alpha \phi_1 + \alpha_2 \phi_2.$$

Then, by orthogonality, the l_2 norm is calculable from either set of coordinates (x_1, x_2) or (α, β) the same way

$$\|x\|^2 = x_1^2 + x_2^2 = \alpha^2 + \beta^2.$$

On the other hand, consider the A norm:

$$\|x\|_A^2 = (\alpha_1\phi_1 + \beta\phi_2)^t \, A \, (\alpha_1\phi_1 + \beta\phi_2)$$
$$= (\alpha_1\phi_1 + \beta\phi_2)^t \, A \, (\alpha_1\lambda_1\phi_1 + \beta\lambda_2\phi_2)$$
$$\text{by orthogonality of } \phi_1, \phi_2$$
$$= \lambda_1\alpha^2 + \lambda_2\beta^2.$$

Comparing the ℓ_2 norm and the A norm

$$\|x\|^2 = \alpha^2 + \beta^2 \quad and \quad \|x\|_A^2 = \lambda_1\alpha^2 + \lambda_2\beta^2,$$

we see that $\| \cdot \|_A$ is again exactly a weighted ℓ_2 norm, weighted by the eigenvalues of A.

Exercise 7.1. *For A either (i) not symmetric, or (ii) indefinite, consider*

$$x, y \rightarrow \langle x, y \rangle_A.$$

In each case, what properties of an inner product fail?

Exercise 7.2. *If A is skew symmetric show that for real vectors $x^t A x = 0$. Given an $N \times N$ matrix A, split A into its symmetric and skew-symmetric parts by*

$$A_{symmetric} = \frac{A + A^t}{2}$$

$$A_{skew} = \frac{A - A^t}{2}.$$

Verify that $A = A_{symmetric} + A_{skew}$. Use this splitting to show that any positive real matrix is the sum of an SPD matrix and a skew symmetric matrix.

7.1 The Connection to Optimization

"Nature uses as little as possible of anything."
— Kepler, Johannes (1571–1630)

We consider the solution of $Ax = b$ for SPD matrices A. This system has a deep connection to an associated optimization problem. For A an $N \times N$ SPD matrix, define the function $J(x_1, \ldots, x_N) \to \mathbb{R}$ by

$$J(x) = \frac{1}{2} x^t Ax - x^t b.$$

Theorem 7.1. *Let A be SPD. The solution of $Ax = b$ is the unique minimizer of $J(x)$. There holds*

$$J(x + y) = J(x) + \frac{1}{2} y^t Ay > J(x) \text{ for any } y.$$

Further, if $\widehat{x}$ is any other vector in $\mathbb{R}^N$ then

$$\|x - \widehat{x}\|_A^2 = 2\left(J(\widehat{x}) - J(x)\right). \tag{7.1}$$

Proof. This is an identity. First note that since A is SPD if $y \neq 0$ then $y^t Ay > 0$. We use $x = A^{-1}b$ and A is SPD. Expand and collect terms:

$$
\begin{aligned}
J(x + y) &= \frac{1}{2}(x+y)^t A(x+y) - (x+y)^t b \\
&= \frac{1}{2} x^t Ax + \frac{1}{2} x^t Ax + \frac{1}{2} \cdot 2y^t Ax + \frac{1}{2} y^t Ay - x^t b - y^t b \\
&= J(x) + y^t(Ax - b) + \frac{1}{2} y^t Ay + \frac{1}{2} y^t Ay \\
J(x) &+ \frac{1}{2} y^t Ay > J(x).
\end{aligned}
$$

This is the first claim. The second claim is also an identity: we expand the LHS and RHS and cancel terms until we reach something equal. The formal proof is then this verification in reverse. Indeed, expanding

$$
\begin{aligned}
\|x - \widehat{x}\|_A^2 &= (x - \widehat{x})^t A(x - \widehat{x}) = x^t Ax - \widehat{x}^t Ax - x^t A\widehat{x} + \widehat{x}^t A\widehat{x} \\
&= (\text{since } Ax = b) = x^b - \widehat{x}^t b - \widehat{x}^t b + \widehat{x}^t A\widehat{x} = \widehat{x}^t A\widehat{x} - 2\widehat{x}^t b + x^t b
\end{aligned}
$$

and

$$
\begin{aligned}
2\left(J(\widehat{x}) - J(x)\right) &= \widehat{x}^t A\widehat{x} - 2\widehat{x}^t b - x^t Ax + 2x^t b \\
&= (\text{since } Ax = b) = \widehat{x}^t A\widehat{x} - 2\widehat{x}^t b - x^t[Ax - b] + x^t b \\
&= \widehat{x}^t A\widehat{x} - 2\widehat{x}^t b + x^t b,
\end{aligned}
$$

which are obviously equal. Each step is reversible so the result is proven. $\square$

Thus, for SPD A we can write.

Corollary 7.1. *For A SPD the following problems are equivalent:*

$$solve : Ax = b,$$

$$minimize : J(y).$$

The equivalence can be written using the terminology of optimization as $x = A^{-1}b$ is the **arg**ument that **min**imizes $J(y)$:

$$x = \arg \min_{y \in \mathbb{R}^N} J(y).$$

Example 7.3 (The 2×2 case). Consider the 2×2 linear system $A\vec{x} = \vec{b}$. Let A be the symmetric 2×2 matrix

$$A = \begin{bmatrix} a & c \\ c & d \end{bmatrix}.$$

Calculating the eigenvalues, it is easy to check that A is SPD if and only if

$$a > 0, d > 0, \text{ and } c^2 - ad < 0.$$

Consider the energy functional $J(x)$. Since $\vec{x}$ is a 2 vector, denote it by $(x, y)^t$. Since the range is scalar, $z = J(x, y)$ is an energy surface:

$$z = J(x, y) = \frac{1}{2}(x, y) \begin{bmatrix} a & c \\ c & d \end{bmatrix} \begin{pmatrix} x \\ y \end{pmatrix} - (x, y) \begin{pmatrix} b_1 \\ b_2 \end{pmatrix}$$

$$= \frac{1}{2} \left[ax^2 + 2cxy + dy^2 \right] - (b_1 x + b_2 y).$$

This surface, shown in Figure 7.1, is a paraboloid opening up if and only is the above condition on the eigenvalues hold: $a > 0$, $d > 0$, and $c^2 - ad < 0$. One example is plotted below. The solution of $Ax = b$ is the point in the $x - y$ plane where $z = J(x, y)$ attains its minimum value.

Minimization problems have the added advantage that it is easy to calculate if an approximate solution has been improved: If J(the new value) < J(the old value) then it has! It is important that the amount $J(*)$ decreases correlates exactly with the decrease in the A-norm error as follows.

Equation (7.1) shows clearly that solving $Ax = b$ (so $\hat{x} = x$) is equivalent to minimizing $J(\cdot)$ (since $J(\hat{x}) \geq J(x)$ and equals $J(x)$ only when $\hat{x} \equiv x$). Theorem 7.1 and Corollary 7.1 show that powerful tools from optimization can be used to solve $Ax = b$ when A is SPD. There is a wide class of iterative methods from optimization that take advantage of this equivalence: descent methods. The prototypical descent method is as follows.

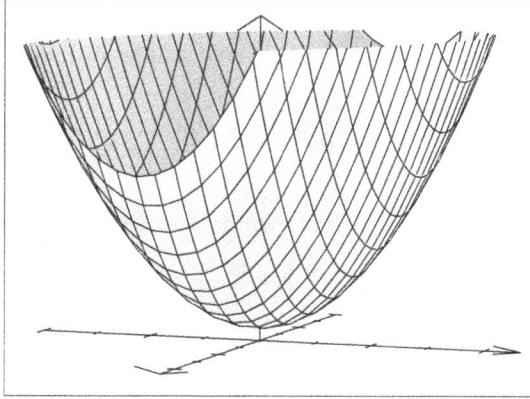

Fig. 7.1   An example of $z = J(x, y)$.

Algorithm 7.1 (General descent method). *Given* $Ax = b$, *a quadratic functional* $J(\cdot)$ *that is minimized at* $x = A^{-1}b$, *a maximum number of iterations* itmax *and an initial guess* x^0:

> *Compute* $r^0 = b - Ax^0$
> for n=1:itmax
> $(*)$ *Choose a direction vector* d^n
>   *Find* $\alpha = \alpha_n$ *by solving the 1D minimization problem:*
> $(**)$ $\alpha_n = \arg\min_\alpha \Phi(x^n + \alpha d^n)$
> $x^{n+1} = x^n + \alpha_n d^n$
> $r^{n+1} = b - Ax^{n+1}$
>   if *converged, exit,* end
> end

The most common examples of steps $(*)$ and $(**)$ are:

- Functional: $\Phi(x) := J(x) = \frac{1}{2}x^t A x - x^t b$,
- Descent direction: $d^n = -\nabla J(x^n)$.

These choices yield the steepest descent method. Because the functional $J(x)$ is quadratic, there is a very simple formula for α_n in step $(**)$ for steepest descent:

$$\alpha_n = \frac{d^n \cdot r^n}{d^n \cdot A d^n} \text{ , where } r^n = b - Ax^n. \tag{7.2}$$

It will be convenient to use the $\langle \cdot, \cdot \rangle$ notation for dot products so this formula is equivalently written

$$\alpha_n = \frac{\langle d^n, r^n \rangle}{\langle d^n, Ad^n \rangle} = \frac{\langle d^n, r^n \rangle}{\langle d^n, d^n \rangle_A}.$$

The difference between descent methods arises from:

(1) The functional minimized, and most commonly,
(2) The choice of descent direction.

Many choices of descent direction and functionals have been tried. Examples of other choices include the following:

Choice of descent direction d^n:

- Steepest descent direction: $d^n = -\nabla J(x^n)$.
- Random directions: $d^n =$ a randomly chosen vector
- Gauss–Seidel like descent: d^n cycles through the standard basis of unit vectors $e_1, e_2, \cdots, e_N$ and repeats if necessary.
- Conjugate directions: d^n cycles through an A-orthogonal set of vectors.

Choice of Functionals to minimize:

- If A is SPD the most common choice is

$$J(x) = \frac{1}{2} x^t A x - x^t b.$$

- Minimum residual methods: for general A,

$$J(x) := \frac{1}{2} \langle b - Ax, b - Ax \rangle.$$

- Various combinations such as residuals plus updates:

$$J(x) := \frac{1}{2} \langle b - Ax, b - Ax \rangle + \frac{1}{2} \langle x - x^n, x - x^n \rangle.$$

Exercise 7.3. *Prove (7.2) that* $\alpha_n = \frac{d^n \cdot r^n}{d^n \cdot Ad^n}$. *Hint: Set* $\frac{d}{d\alpha}(J(x + \alpha d)) = 0$ *and solve.*

Exercise 7.4. *Consider solving $Ax = b$ by a descent method for a general non-SPD, matrix A. Rewrite the above descent algorithm to minimize at each step $||b - Ax^n||_2^2 := r^n \cdot r^n$. Find a formula for α_n. Find the steepest descent direction for $||b - Ax^n||_2^2$.*

Exercise 7.5. *If A is an $N \times N$ SPD matrix and one has access to a complete set of A-orthogonal vectors $\phi_1, \cdots, \phi_N$ show that the solution to $Ax = b$ can be written down in closed form (but using inner products). Find the number of FLOPs required to get the solution by just calculating the closed form solution.*

Exercise 7.6. *For A SPD and C an $N \times N$ matrix and ε a small parameter, consider the minimization problem:*

$$x_\varepsilon = \arg\min J_\varepsilon(x) := \frac{1}{2}x^t Ax + \frac{1}{2\varepsilon}\|Cx\|_2^2 - x^t b.$$

Find the linear system x_ε satisfies. Prove the coefficient matrix is SPD. Show that $x_\varepsilon \to A^{-1}b$ as $\varepsilon \to \infty$. Next consider the case $\varepsilon \to 0$ and show

$$x_\varepsilon \to Nullspace(C), \quad i.e., \quad Cx_\varepsilon \to 0.$$

Exercise 7.7. *Let A be the symmetric 2×2 matrix*

$$A = \begin{bmatrix} a & c \\ c & d \end{bmatrix}.$$

Find a necessary and sufficient condition on $trace(A)$ and $det(A)$ for A to be SPD.

7.2 Application to Stationary Iterative Methods

> "As a historian, I cannot believe how low the standards are in mathematics! In my field, no one would put forth an argument without at least ten proofs, whereas in mathematics they stop as soon as they have found a single one!"
> — An irate historian berating Andrey Kolmogorov.
> "How long will you delay to be wise?" — Epictetus

Consider a stationary iterative method based on decomposing A by $A = M - N$. With this additive decomposition, $Ax = b$ is equivalent to $Mx = b + Nx$. The induced iterative method is then

$$M(x_{n+1} - x_n) = b - Ax_n \quad \text{or} \qquad (7.3)$$
$$Mx_{n+1} = b + Nx_n.$$

Obviously, if $M = A$ this converges in one step but that one step is just solving $Ax = b$. The matrix M must approximate A and yet systems $Mx^{n+1} = RHS$ must also be very easy to solve. Sometimes such a matrix M is call a *preconditioner* of the matrix A and $A = M - N$ is often called a *regular splitting* of A. Examples include

- FOR: $M = \rho I$, $N = \rho I - A$
- Jacobi: $M = diag(A)$
- Gauss–Seidel: $M = D + L$ (the lower triangular part of A).

Householder (an early giant in numerical linear algebra and matrix theory) proved a very simple identity for (7.3) when A is SPD.

Lemma 7.2 (Householder lemma). *Let A be SPD and let x_n be given by (7.3). With $e_n = x - x_n$*

$$e_n^t A e_n - e_{n+1}^t A e_{n+1} = \frac{1}{2}(x_{n+1} - x_n)P(x_{n+1} - x_n) \qquad (7.4)$$

where $P = M + M^t - A$.

Proof. This is an identity: expand both sides and cancel to check that is true. Next reverse the steps to give the proof. ☐

Corollary 7.2 (Convergence of FOR, Jacobi, GS and SOR).
For A SPD, if P is positive definite (7.3) converges. The convergence is monotonic in the A norm:

$$\|e_n\|_A > \|e_{n+1}\|_A > \ldots \longrightarrow 0 \quad as \quad n \to 0.$$

Proof. The proof is easy but there are so many tools at hand it is also easy to start on the wrong track and get stuck there. Note that $\|e_n\|_A$ is monotone decreasing and bounded below by zero. Thus it has a non-negative limit. Since it converges to something, the Cauchy criteria implies that

$$\left(e_n^t A e_n - e_{n+1}^t A e_{n+1}\right) \to 0.$$

Now reconsider the Householder Relation (7.4). Since the LHS $\to 0$ we must have the RHS $\to 0$ too.[1] Since $P > 0$, this means

$$\|x_{n+1} - x_n\|_P \to 0.$$

Finally the iteration itself,

$$M\left(x^{n+1} - x^n\right) = b - A x^n,$$

implies that if $x_{n+1} - x_n \to 0$ (the LHS), then the RHS does also: $b - A x^n \to 0$, so convergence follows. ☐

---

[1]This step is interesting to the study of human errors. Since we spend our lifetime reading and writing L to R, top to bottom, it is common for our eyes and brain to process the mathematics = sign as a one directional relation ⇒ when we are in the middle of a proof attempt.

Let us apply this result to the above examples.

Theorem 7.2 (Convergence of FOR, GS, Jacobi, SOR).

- *FOR converges monotonically in* $\|\cdot\|_A$ *if* [2]

$$P = M + M^t - A = 2\rho I - A > 0 \quad \text{if} \ \rho > \frac{1}{2}\lambda_{\max}(A).$$

- *Jacobi converges monotonically in the A-norm if* $diag(A) > \frac{1}{2}A$.
- *Gauss–Seidel converges monotonically in the A norm for SPD A in all cases.*
- *SOR converges monotonically in the A-norm if* $0 < \omega < 2$.

Proof. This follows easily from Householder's result as follows. For Jacobi, $M = diag(A)$, so

$$P = M + M^t - A = 2diag(A) - A > 0 \text{ if } diag(A) > \frac{1}{2}A.$$

For GS, since $A = D + L + U$ where (since A is symmetric) $U = L^t$ and

$$P = M + M^t - A = (D + L) + (D + L)^t - A =$$
$$D + L + D + L^t - (D + L + L^t) = D > 0$$

for SPD A. For SOR we calculate (as above using $L^t = U$)

$$P = M + M^t - A = M + M^t - (M - N) = M^t + N$$
$$= \omega^{-1}D + L^t + \frac{1-\omega}{\omega}D - U = \frac{2-\omega}{\omega}D > 0,$$

for $0 < \omega < 2$. Convergence of FOR in the A-norm is left as an exercise. □

"WRITE. FINISH THINGS. KEEP WRITING."
— Neil Gaiman

Exercise 7.8. *Consider the 2D MPP on a uniform $N \times N$ mesh. Divide the domain in half $\Omega = \Omega_1 \cup \Omega_2$ (not through any meshpoints) partitioning the mesh into two subsets of equal numbers. This then partitions the solution and RHS accordingly as (if we first order the mesh points in Ω_1 then in Ω_2) $u = (u_1, u_2)$. Show that the MPP then takes the block form*

$$\begin{bmatrix} A_1 & -C \\ -C & A_2 \end{bmatrix} \begin{bmatrix} u_1 \\ u_2 \end{bmatrix} = \begin{bmatrix} f_1 \\ f_2 \end{bmatrix}.$$

---

[2]Here $A > B$ means $(A - B)$ is positive definite, i.e. $x^t A x > x^t B x$ for all $x \neq 0$. Also, monotonic convergence in the A-norm means the errors satisfy $\|e^{n+1}\|_A < \|e^n\|_A$ for all n.

Find the form of A_1 and A_2. Show that they are diagonally semi-dominant. Look up the definition and show they are also irreducibly diagonally dominant. Show that the entries in C are nonnegative.

Exercise 7.9 (Convergence of block Jacobi). Continuing the last problem, consider the block Jacobi method given below

$$\begin{bmatrix} A_1 & 0 \\ 0 & A_2 \end{bmatrix} \left(\begin{bmatrix} u_1^{n+1} \\ u_2^{n+1} \end{bmatrix} - \begin{bmatrix} u_1^n \\ u_2^n \end{bmatrix} \right) = \begin{bmatrix} f_1 \\ f_2 \end{bmatrix} - \begin{bmatrix} A_1 & -C \\ -C & A_2 \end{bmatrix} \begin{bmatrix} u_1^n \\ u_2^n \end{bmatrix}.$$

Use Householders theorem to prove this converges.

Exercise 7.10. Repeat the above for block FOR and for Block Gauss–Seidel.

Exercise 7.11 (Red-Black block methods). Consider the $2D$ MPP on a uniform $N \times N$ mesh. Draw a representative mesh and color the meshpoints by red-black like a typical checkerboard (chess players should think of greed and buff). Note that the 5 point star stencil links red points only to black and black only to red. Order the unknowns as first red then black, partitioning the mesh vertices into two subsets of about equal numbers. This then partitions the solution and RHS accordingly as $u = (u_{RED}, u_{BLACK})$. Show that the MPP then takes the block form

$$\begin{bmatrix} A_1 & -C \\ -C & A_2 \end{bmatrix} \begin{bmatrix} u_{RED} \\ u_{BLACK} \end{bmatrix} = \begin{bmatrix} f_{RED} \\ f_{BLACK} \end{bmatrix}.$$

Find the form of $A_{1,2}$. It will be best to do this for a fixed, e.g., 4×4 mesh before jumping to the general mesh. Analyze the structure of the submatrices. Based on their structure, propose and analyze convergence of a block iterative method. Again, try it on a 4×4 mesh first.

Exercise 7.12. Let A be $N \times N$ and SPD. Consider FOR for solving $Ax = b$. Show that for optimal ρ we have

$$J(x^n) - J(x) \leq \left(\frac{\lambda_{\max}(A) - \lambda_{\min}(A)}{\lambda_{\max}(A) + \lambda_{\min}(A)} \right) \left(J(x^{n-1}) - J(x) \right).$$

Express the multiplier $\frac{\lambda_{\max} - \lambda_{\min}}{\lambda_{\max} + \lambda_{\min}}$ in terms of $\text{cond}_2(A)$.

Exercise 7.13. Consider the proof of convergence when $P > 0$. This proof goes back and forth between the minimization structure of the iteration and the algebraic form of it. Try to rewrite the proof entirely in terms of the functional $J(x)$ and $\nabla J(x)$.

Exercise 7.14. *Give a complete and detailed proof of the Householder lemma. Give the details of the proof that $P > 0$ implies convergence.*

7.3 Application to Parameter Selection

> "The NSA is a self-licking ice cream cone."
> — An anonymous senior official of the National Security Agency.

Consider Richardson's method FOR for A an SPD matrix:

$$\rho(x^{n+1} - x^n) = b - Ax^n, \qquad \text{or} \qquad x^{n+1} = x^n + \rho^{-1}r^n$$

where $r^n = b - Ax^n$. We have an idea of "optimal" value of ρ

$$\rho_{\text{optimal}} = (\lambda_{\max}(A) + \lambda_{\min}(A))/2$$

which minimizes the maximum error over all possible initial conditions. It is, alas, hard to compute.

We consider here another idea of optimal:

Given x^n, find $\rho = \rho_n$ which will make x^{n+1} as accurate as possible on that step.

The algorithm would be:

Algorithm 7.2. *Given x^0 and a maximum number of iterations,* itmax*:*

```
for n=0:itmax
```
    Compute $r^n = b - Ax^n$
    Compute ρ^n via a few auxiliary calculations
    $x^{n+1} = x^n + (\rho^n)^{-1}r^n$
    if converged, exit, **end**
end

Lemma 7.3. *In exact arithmetic, the residuals $r^n = b - Ax^n$ of FOR satisfies*

$$r^{n+1} = r^n - \rho^{-1}Ar^n.$$

Proof.

$$x^{n+1} = x^n - \rho^{-1}r^n.$$

Multiply by "$-A$" and add b to both sides. This gives

$$b - Ax^{n+1} = b - Ax^n - \rho^{-1}Ar^n,$$

which is the claimed iteration. $\qquad\square$

Exercise 7.15. *In Chapter 6, Exercise 6.4 (page 120) required a computer program to FOR for the 2D MPP with $\rho = 4$ (the Jacobi method). Modify this computer program so that it can use an arbitrary ρ. The 2D analog of Theorem 5.3 (page 110) in Chapter 5 shows that $\lambda_{max} \doteq 8$ and $\lambda_{min} \doteq h^2$, so a reasonably good choice is $\rho = (8 + h^2)/2$.*

Test your program by solving the 2D MPP with $h = 1/10$, RHS $f = 0$, and with boundary conditions $g(x, y) = x - y$. Use as initial guess the exact solution $u = x - y$. You should observe convergence in a single iteration. If it takes more than five iterations, or if it does not converge, you have an error in your program.

How many iterations are required to reach a convergence criterion of $1.e - 4$ when $h = 1/100$ and the initial guess is $u(x, y) = 0$ in the interior and $u(x, y) = x - y$ on the boundary?

Different formulas for selecting ρ emerge from different interpretations of what "as accurate as possible" means.

Option 1: Residual minimization: Pick ρ^n to minimize $\|r^{n+1}\|^2$. By the last lemma,

$$\|r^{n+1}\|^2 = \|r^n - \rho^{-1}Ar^n\|^2 = \langle r^n - \rho^{-1}Ar^n, r^n - \rho^{-1}Ar^n \rangle.$$

Since r^n is fixed, this is a simple function of ρ

$$\widetilde{J}(\rho) = \langle r^n - \rho^{-1}Ar^n, r^n - \rho^{-1}Ar^n \rangle$$

or

$$\widetilde{J}(\rho) = \langle r^n, r^n \rangle - 2\langle r^n, Ar^n \rangle \rho^{-1} - \langle Ar^n, Ar^n \rangle \rho^{-2}.$$

Taking $\widetilde{J}'(\rho) = 0$ and solving for $\rho = \rho_{\text{optimal}}$ gives

$$\rho_{\text{optimal}} = \frac{\langle Ar^n, Ar^n \rangle}{\langle r^n, Ar^n \rangle}.$$

The cost of using this optimal value at each step: two extra dot products per step.

Option 2: J minimization: Pick ρ^n to minimize $J(x^{n+1})$. In this case we define

$$\phi(\rho) = J(x^{n+1}) = J(x^n + \rho^{-1}r^n)$$

$$= \frac{1}{2}(x^n + \rho^{-1}r^n)^t A(x^n + \rho^{-1}r^n) - (x^n + \rho^{-1}r^n)^t b.$$

Expanding, setting $\phi'(\rho) = 0$ and solving, as before, gives

$$\rho_{\text{optimal}} = \frac{\langle r^n, Ar^n \rangle}{\langle r^n, r^n \rangle}.$$

Option 2 is only available for SPD A. However, for such A it is preferable to Option 1. It gives the algorithm

Algorithm 7.3. *Given x^0 the matrix A and a maximum number of iterations,* itmax*:*

$$r^1 = b - Ax^1$$
for n=1:itmax
$$\rho^n = \frac{\langle Ar^n, r^n \rangle}{\langle r^n, r^n \rangle}$$
$$x^{n+1} = x^n + (\rho^n)^{-1} r^n$$
if *satisfied, exit,* end
$$r^{n+1} = b - Ax^{n+1}$$
end

Exercise 7.16. *In Exercise 7.15 you wrote a computer program to use FOR for the 2D MPP with arbitrary ρ, defaulting to the optimal value for the 2D MPP. In this exercise, you will modify that program to make two other programs: (a) One for Algorithm 7.3 for Option 2, and, (b) One for Option 1.*

In each of these cases, test your program by solving the 2D MPP with $h = 1/10$, RHS $f = 0$, and with boundary conditions $g(x, y) = x - y$. Use as initial guess the exact solution $u = x - y$. You should observe convergence in a single iteration. If it takes more than five iterations, or if it does not converge, you have an error in your program.

To implement Algorithm 7.3, you will have to write code defining the vector variable r *for the residual r^n and in order to compute the matrix-vector product Ar^n, you will have write code similar to the code for* au *(giving the product Au), but defining a vector variable. This is best done in a separate loop from the existing loop.*

To implement Option 1, the residual minimization option, all you need to do is change the expression for ρ.

In each case, how many iterations are required for convergence when $h = 1/100$ when the initial guess is $u(x, y) = 0$ in the interior and $u(x, y) = x - y$ on the boundary?

The connection between Options 1 and 2 is through the (celebrated) "normal" equations. Since $Ae = r$, minimizing $||r||_2^2 = r^t r$ is equivalent to minimizing $||Ae||_2^2 = e^t A^t Ae = ||e||_{A^t A}^2$. Since $A^t A$ is SPD minimizing $||e||_{A^t A}^2$ is equivalent to minimizing the quadratic functional associated with

$(A^t A)x = A^t b.$

$$\widetilde{J}(x) = \frac{1}{2} x^t A^t A x - x^t A^t b.$$

If we are solving $Ax = b$ with A an $N \times N$ nonsingular matrix then we can convert it to the normal equations by multiplication by A^t:

$$(A^t A)x = A^t b.$$

Thus, minimizing the residual is equivalent to passing to the normal equations and minimizing $J(\cdot)$. Unfortunately, the bandwidth of $A^t A$ is (typically) double the bandwidth of A. Further, passing to the normal equations squares the condition number of the associated linear system.

Theorem 7.3 (The normal equations). *Let A be $N \times N$ and invertible. Then $A^t A$ is SPD. If A is SPD then $\lambda(A^t A) = \lambda(A)^2$ and*

$$cond_2(A^t A) = [cond_2(A)]^2 .$$

Proof. Symmetry: $(A^t A)^t = A^t A^{tt} = A^t A$. Positivity: $x^t(A^t A)x = (Ax)^t Ax = |Ax|^2 > 0$ for x nonzero since A is invertible. If A is SPD, then $A^t A = A^2$ and, by the spectral mapping theorem,

$$cond_2(A^t A) = cond_2(A^2) = \frac{\lambda_{\max}(A^2)}{\lambda_{\min}(A^2)}$$

$$= \frac{\lambda_{\max}(A)^2}{\lambda_{\min}(A)^2} = \left(\frac{\lambda_{\max}(A)}{\lambda_{\min}(A)} \right)^2$$

$$= [cond_2(A)]^2 .$$

$\square$

The relation $cond_2(A^t A) = [cond_2(A)]^2$ explains why Option 2 is better. Option 1 implicitly converts the system to the normal equations and thus squares the condition number of the system being solved then applies Option 2. This results in a very large increase in the number of iterations.

7.4 The Steepest Descent Method

A small error in the former will produce an enormous error in the latter.
— Henri Poincaré

We follow two rules in the matter of optimization:

Rule 1. Don't do it.

Rule 2 (for experts only). Don't do it yet - that is, not until you have a perfectly clear and unoptimized solution.

— M. A. Jackson

"Libenter homines et id quod volunt credunt." — an old saying.

The steepest descent method is an algorithm for minimizing a functional in which, at each step, the choice of descent direction is made which makes the functional decrease as much as possible at that step. Suppose that a functional $J(x)$ is given. The direction in which $J(\cdot)$ decreases most rapidly at a point x^n is

$$(*) \quad d = -\nabla J(x^n).$$

Consider the line L in direction d passing through x^n. For $\alpha \in \mathbb{R}$, L is given by the equation

$$x = x^n + \alpha d.$$

Steepest descent involves choosing α so that $J(\cdot)$ is maximally decreased on L,

$$(**) \quad J(x^n + \alpha_n d) = \min_{\alpha \in \mathbb{R}} J(x^n + \alpha d).$$

When A is an SPD matrix and $J(x) = \frac{1}{2}x^t A x - x^t b$ each step can be written down explicitly. For example, simple calculations show

$$d^n := -\nabla J(x^n) = r^n = b - Ax^n,$$

and for α_n we solve for $\alpha = \alpha_n$ in

$$\frac{d}{d\alpha} J(x^n + \alpha d^n) = 0$$

to give α_n

$$\alpha_n = \frac{\langle d^n, r^n \rangle}{\langle d^n, Ad^n \rangle}, \quad \text{in general,}$$

and with $d^n = r^n$:

$$\alpha_n = \frac{\langle r^n, r^n \rangle}{\langle r^n, Ar^n \rangle} = \frac{||r^n||^2}{||r^n||_A^2}.$$

Algorithm 7.4 (Steepest descent). *Given an SPD A, x^0, $r^0 = b - Ax^0$ and a maximum number of iterations* `itmax`

```
for n=0:itmax
    rⁿ = b - Axⁿ
    αₙ = ⟨rⁿ, rⁿ⟩/⟨rⁿ, Arⁿ⟩
    xⁿ⁺¹ = xⁿ + αₙrⁿ
    if converged, exit, end
end
```

Comparing the above with FOR with "optimal" parameter selection we see that Steepest descent (Algorithm 7.3 corresponding to Option 2) is exactly FOR with $\alpha^n = 1/\rho^n$ where ρ^n is picked to minimize $J(\cdot)$ at each step.

How does it really work? It gives only marginal improvement over constant α. We conclude that better search directions are needed. The next example shows graphically why Steepest Descent can be so slow:

Example 7.4. $N = 2$, i.e., $\vec{x} = (x, y)^t$. Let $A = \begin{bmatrix} 2 & 0 \\ 0 & 50 \end{bmatrix}$, $b = \begin{bmatrix} 2 \\ 0 \end{bmatrix}$. Then

$$J(\vec{x}) = \frac{1}{2}[x, y]\begin{bmatrix} 2 & 0 \\ 0 & 50 \end{bmatrix}\begin{bmatrix} x \\ y \end{bmatrix} - [x, y]\begin{bmatrix} 2 \\ 0 \end{bmatrix}$$

$$= \frac{1}{2}(2x^2 + 50y^2) - 2x = \underbrace{x^2 - 2x + 25y^2}_{\text{ellipse}} + 1 - 1$$

$$= \frac{(x-1)^2}{1^2} + \frac{y^2}{(\frac{1}{5})^2} - 1.$$

Convergence of Steepest Descent

The fundamental convergence theorem of steepest descent (given next) asserts a worst case rate of convergence that is no better than that of FOR. Unfortunately, the predicted rate of convergence is sharp.

Theorem 7.4 (Convergence of SD). *Let A be SPD and $\kappa = \lambda_{\max}(A)/\lambda_{\min}(A)$. The steepest descent method converges to the solution of $Ax = b$ for any x^0. The error $x - x^n$ satisfies*

$$\|x - x^n\|_A \leq \left(\frac{\kappa - 1}{\kappa + 1}\right)^n \|x - x^0\|_A$$

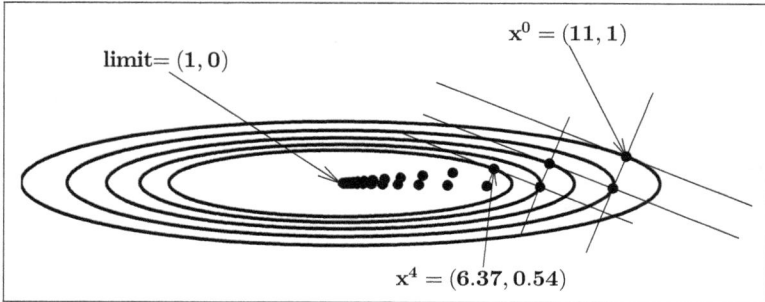

Fig. 7.2   The first minimization steps for Example 7.4. The points $x^0, \ldots, x^4, \ldots$ are indicated with dots, the level curves of J are ellipses centered at $(1,0)$ and construction lines indicate search directions and tangents.

and

$$J(x^n) - J(x) \le \left(\frac{\kappa - 1}{\kappa + 1}\right)^n \left(J(x^0) - J(x)\right).$$

Proof. We shall give a short proof that for one step of steepest descent

$$\|x - x^n\|_A \le \left(\frac{\kappa - 1}{\kappa + 1}\right)\|x - x^{n-1}\|_A.$$

If this holds for one step then the claimed result follows for n steps. We observe that this result has already been proven! Indeed, since steepest descent picks ρ to reduce $J(\cdot)$ *maximally* and thus the A norm of the error maximally going from x^{n-1} to x^n it must also reduce it more than for any other choice of ρ including $\rho_{optimal}$ for FOR. Let x^n_{FOR} be the result from one step from x^{n-1} of First Order Richardson with optimal parameter. We have proven that

$$\|x - x^n_{FOR}\|_A \le \left(\frac{\kappa - 1}{\kappa + 1}\right)\|x - x^{n-1}\|_A.$$

Thus

$$\|x - x^n\|_A \le \|x - x^n_{FOR}\|_A \le \left(\frac{\kappa - 1}{\kappa + 1}\right)\|x - x^{n-1}\|_A,$$

completing the proof for the error. The second result for $J(x^n) - J(x)$ is left as an exercise.   □

We note that $\frac{\kappa-1}{\kappa+1} = 1 - \frac{2}{\kappa+1}$. For the model Poisson problem, typically $\lambda_{\max} = O(1)$ while $\lambda_{\min} = O(h^2)$ and thus $\kappa = O(h^2)$ so steepest descent requires $O(h^{-2})$ directions to converge.

Theorem 7.5. *The convergence rate $\frac{\kappa-1}{\kappa+1}$ of steepest descent is sharp. It is exactly the rate of convergence when the initial error is $e^0 = \phi_1 + \phi_2$ and when $e^0 = \phi_1 - \phi_2$ where $\phi_{1,2}$ are the eigenvectors of $\lambda_{\min}(A)$ and $\lambda_{\max}(A)$ respectively.*

Proof. Let $\phi_{1,2}$ be the eigenvectors of $\lambda_{\min}(A)$ and $\lambda_{\max}(A)$ respectively. Consider two possible selections of initial guesses: Pick

$$x^0 = x - (\phi_1 + \phi_2) \qquad or \qquad x^0 = x - (\phi_1 - \phi_2).$$

We proceed by direct calculations (which are not short but routine step by step): if we choose $x^0 = x + (\phi_1 + \phi_2)$ then $e^0 = \phi_1 + \phi_2$. We find

$$x^1 = x^0 + \alpha_0(b - Ax^0) =$$
$$x^1 = x^0 + \alpha_0 Ae^0 \qquad \text{(since } Ae = r)$$
$$x - x^1 = x - x^0 - \alpha_0 Ae^0$$
$$e^1 = e^0 - \alpha_0 Ae^0 = (\phi_1 + \phi_2) - \alpha_0 A(\phi_1 + \phi_2)$$
$$e^1 = (1 - \alpha_0 \lambda_{\min})\phi_1 + (1 - \alpha_0 \lambda \max)\phi_2.$$

Next calculate similarly

$$\alpha_0 = \frac{\langle r^0, r^0 \rangle}{\langle r^0, Ar^0 \rangle}$$
$$= \frac{\langle Ae^0, Ae^0 \rangle}{\langle Ae^0, A^2 e^0 \rangle}$$
$$= \frac{\langle A(\phi_1 + \phi_2), A(\phi_1 + \phi_2) \rangle}{\langle A(\phi_1 + \phi_2), A^2(\phi_1 + \phi_2) \rangle}$$
$$= \cdots = \frac{2}{\lambda_{\min} + \lambda_{\max}}.$$

We thus have

$$e^1 = \left(1 - \frac{2\lambda_{\min}}{\lambda_{\min} + \lambda_{\max}}\right)\phi_1 + (1 - \frac{2\lambda_{\max}}{\lambda_{\min} + \lambda_{\max}})\phi_2$$
$$= (rearranging) = \left(\frac{\kappa - 1}{\kappa + 1}\right)(\phi_1 - \phi_2).$$

The rest of the calculations are exactly as above. These show that in the two cases

$$e^1 = \left(\frac{\kappa - 1}{\kappa + 1}\right)(\phi_1 \mp \phi_2)$$
$$e^2 = \left(\frac{\kappa - 1}{\kappa + 1}\right)^2 (\phi_1 \pm \phi_2).$$

Proceeding by induction,

$$e^n = \left(\frac{\kappa - 1}{\kappa + 1}\right)^n (\phi_1 \text{ either } \pm \text{ or } \mp \phi_2),$$

in the two cases, which is exactly the predicted rate of convergence.   □

Exercise 7.17. *Suppose you must solve a very large sparse linear system* $Ax = b$ *by some iterative method. Often one does not care about the individual millions of entries in the solution vector but one only wants a few statistics [i.e., numbers] such as the average. Obviously, the error in the averages can be much smaller than the total error in every component or just as large as the total error. Your goal is to try to design iterative methods which will produce accurate statistics more quickly than an accurate answer.*

To make this into a math problem, let the (to fix ideas) statistic be a linear functional of the solution. Define a vector l *and compute*

$$L = l^t x = \langle l, x \rangle$$

$$\text{if, e.g., } L = average(x) \text{ then}$$

$$l = (1/N, 1/N, ..., 1/N)^t.$$

Problem:

$$Solve : Ax = b,$$

$$Compute : L = \langle l, x \rangle$$

or: Compute $L = \dots$ *while solving* $Ax = b$ *approximately. There are many iterative methods you have studied. develop/adapt/optimize one [YOUR CHOICE OF IDEAS] for this problem! You must either [YOUR CHOICE] analyze it or give comprehensive numerical tests. Many approaches are possible, e.g., note that this can be written as a* $N + 1 \times N + 1$ *system for* x, L:

$$\begin{bmatrix} A & 0 \\ -l^t & 1 \end{bmatrix} \begin{bmatrix} x \\ L \end{bmatrix} = \begin{bmatrix} b \\ 0 \end{bmatrix}.$$

You will have to **negotiate with this problem** *as well. There is no set answer! Every method can be adapted to compute* L *faster and no method will always be best.*

Exercise 7.18. *The standard test problem for nonsymmetric systems is the* 2D CDEqn = 2D *model discrete Convection Diffusion equation. Here* ε *is a* small to very small positive parameter. (Recall that you have investigated the 1D CDEqn in Exercise 5.5, page 98.)

$$-\varepsilon\Delta u + u_x = f, \quad \text{inside } (0,1) \times (0,1)$$
$$u = g, \quad \text{on the boundary.}$$

Discretize the Laplacian by the usual 5-point star and approximate u_x by

$$u_x(x_I, y_J) \simeq \frac{u(I+1, J) - u(I-1, J)}{2h}.$$

Find the associated difference stencil. This problem has 2 natural parameters:

$$h = \frac{1}{N+1}, \quad \text{the meshwidth; and,}$$

$$Pe := \frac{h}{2\varepsilon}, \quad \text{the "cell Péclet number".}$$

The interesting case is when the cell Péclet[3] number $Pe \gg 1$, i.e., when $\varepsilon \ll h$.

Hint: You have already written programs for the 2D MPP in Exercises 7.15 and 7.16. You can modify one of those programs for this exercise.

(1) Debug your code using $h = 1/5$, $g(x,y) = x - y$, and $f(x,y) = 1$. The exact solution in this case is $u(x,y) = x - y$. Starting from the exact solution, convergence to 1.e-3 should be achieved in a single iteration of a method such as Jacobi (FOR with $\rho = 4$).

(2) Fix $h = 1/50$, $f(x,y) = x + y$, and $g(x,y) = 0$. Pick three iterative methods (your choice). Solve the nonsymmetric linear system for a variety of values[4] of $\varepsilon = 1, 1/10, 1/100, 1/1000, 1/10000$, starting from $u(x,y) = 0$, to an accuracy of 10^{-3}. Report the results, consisting of convergence with the number of iterations or nonconvergence. Describe the winners and losers for small cell Pe and for large cell Pe.

Exercise 7.19. For A a large, sparse matrix and $\|\cdot\|$ the euclidean or l_2 norm, consider a general iterative method below for solving $Ax = b$, starting from a guess vector x^0.

---

[3]The Péclet number is named after the French physicist Jean Claude Eugène Péclet. It is given by Length × Velocity / Diffusion coefficient. In our simple example, the velocity is the vector (1,0) and the diffusion coefficient is ε. The *cell Péclet number,* also denoted by Pe, is the Peclet number associated with one mesh cell so the length is taken to be the meshwidth.

[4]For $\varepsilon = 1$, your solution should appear much like the MPP2D solution with the same right side and boundary conditions. For smaller ε, the peak of the solution is pushed to larger x locations. Nonconvergence is likely for very small ε.

$$r^0 = b - Ax^0$$
```
for n=0:itmax
```
   Choose d^n
   $(*)$ Pick α_n to minimize $\|b - A(x^n + \alpha d^n)\|^2$
   $x^{n+1} = x^n + \alpha_n d^n$
   $(**)$ $r^{n+1} = b - Ax^{n+1}$
   if *converged*, `return`, `end`
```
end
```

(1) **Show that** step $(**)$ can be replaced by: $(**)$ $r^{n+1} = r^n - \alpha_n Ad^n$.
(2) **Find** an explicit formula for the optimal value of α in step $(*)$.

Chapter 8

The Conjugate Gradient Method

"The cook was a good cook,

as cooks go,

and as cooks go,

she went."

— Saki

The conjugate gradient method was proposed by Hestenes and Stiefel in 1952. Initially it was considered a *direct* method for solving $Ax = b$ for A SPD since (in exact arithmetic) it gives the exact solution in N steps or less. Soon it was learned that often a very good solution is obtained after many fewer than N steps. Each step requires a few inner products, and one matrix multiply. Like all iterative methods, its main advantage is when the matrix vector multiply can be done quickly and with minimal storage of A.

8.1   The CG Algorithm

Why not the best? — Jimmy Carter

The conjugate gradient method is the best possible method[1] for solving $Ax = b$ for A an SPD matrix. We thus consider the solution of

$$Ax = b, \text{ where } A \text{ is large, sparse and SPD.}$$

---

[1] "Best possible" has a technical meaning here with equally technical qualifiers. We shall see that the *k*th step of the CG method computes the projection (the *best approximation*) with respect to the A-norm into a k dimensional subspace.

First we recall some notation.

Definition 8.1. Assume that A is an SPD matrix. $\langle x, y \rangle$ denotes the **Euclidean inner product**:

$$\langle x, y \rangle = x^t y = x_1 y_1 + x_2 y_2 + \ldots + x_n y_n.$$

$\langle x, y \rangle_A$ denotes the A-**inner product**

$$\langle x, y \rangle_A = x^t A y = \sum_{i,j=1}^{N} x_i A_{ij} y_j.$$

The A-**norm** is

$$\|x\|_A = \sqrt{\langle x, x \rangle_A} = \sqrt{x^t A x}.$$

The **quadratic functional** associated with $Ax = b$ is

$$J(x) = \frac{1}{2} x^t A x - x^t b.$$

The conjugate gradient method (hereafter: CG) is a descent method. Thus, it takes the general form.

Algorithm 8.1 (Descent Method for solving $Ax = b$ with A SPD).
Given an SPD A, x^0 and a maximum number of iterations itmax

$r^0 = b - Ax^0$
for n=0:itmax
    (∗)*Choose a descent direction d^n*
    $\alpha_n := \arg\min_\alpha J(x^n + \alpha d^n) = \langle d^n, r^n \rangle / \langle d^n, Ad^n \rangle$
    $x^{n+1} = x^n + \alpha_n d^n$
    $r^{n+1} = b - Ax^{n+1}$
    if *converged, stop,* end
end

CG differs from the slow steepest descent method by step (∗) the choice of search directions. In Steepest Descent $d^n = r^n$ while in CG d^n is calculated by a two term recursion that A orthogonalizes the search directions. The CG algorithm is very simple to write down and easy to program. It is given as follows:[2]

---

[2]We shall use fairly standard conventions in descent methods; we will use roman letters with superscripts to denote vectors, $d, r, x, \cdots$, and greek letters, $\alpha, \beta, \cdots$, with subscripts to denote scalars. For example, we denote the nth descent direction vector d^n and the nth scalar multiplier by α_n. One exception is that eigenvectors will commonly be denoted by ϕ.

Algorithm 8.2 (Conjugate Gradient Algorithm). *Given an SPD A,* x^0 *and a maximum number of iterations* `itmax`

$r^0 = b - Ax^0$

$d^0 = r^0$

`for n=1:itmax`

$\quad \alpha_{n-1} = \langle d^{n-1}, r^{n-1} \rangle / \langle d^{n-1}, Ad^{n-1} \rangle$

$\quad x^n = x^{n-1} + \alpha_{n-1} d^{n-1}$

$\quad r^n = b - Ax^n$

$\quad$ `if` *converged, stop,* `end`

$\quad \beta_n = \langle r^n, r^n \rangle / \langle r^{n-1}, r^{n-1} \rangle$

$\quad d^n = r^n + \beta_n d^{n-1}$

`end`

CG has the following features:

- In steepest descent, d^n is chosen to be a locally optimal search direction.
- In CG, d^n is chosen to be a globally optimal search direction. The problem of finding d^n is thus a global problem: in principle, d^n depends on all the previous search directions $d^0, d^1, d^2, \ldots, d^{n-2}$ and d^{n-1}. CG, however, has an amazing property:
- For SPD A, the dependence on $d^0, \ldots, d^{n-3}$ drops out and d^n depends only on d^{n-1} and d^{n-2}.
- CG is the fastest convergent iterative method in the A-norm.
- CG can be written as a three term recursion or a coupled two term recursion.
- CG requires typically $O\left(\sqrt{\text{cond}(A)}\right)$ iterations per significant digit of accuracy.
- CG requires barely more work per step than steepest descent. As stated above, it takes 2 matrix-vector multiplies per step plus a few dot products and triads. If the residual is calculated using Lemma 277 by $r^{n+1} = r^n - \alpha_n Ad^n$, then it only requires one matrix-vector multiply per step.
- In exact arithmetic, CG reaches the exact solution of an $N \times N$ system in N steps or less.
- The CG method has many orthogonality properties. Thus, there are many ways to write the algorithm that are mathematically equivalent (in exact arithmetic) and an apparent (not real) multitude of CG methods.

For general *nonsymmetric* matrices, there is no iterative method with all the above good properties of CG. Two methods that are popular now are **GMRES** which has a full recursion[3] and thus is very expensive when a lot of iterates are required, and **CGN** — which is just CG for the normal equations

$$A^t A x = A^t b, \qquad A^t A \text{ is SPD.}$$

This typically requires $O\left(\text{cond}(A)\right)$ iterates per significant digit of accuracy sought.

Example 8.1. As a concrete example, consider solving the $2D$ model Poisson problem on a 100×100 mesh. Thus $h = \frac{1}{101}$ and we solve

$$A\vec{u} = f \quad \text{where} \quad A \text{ is } 10,000 \times 10,000.$$

Note that $\text{cond}(A) \simeq O(h^{-2}) = O(10,000)$. Thus, we anticipate:

- Steepest descent requires $\simeq 50,000$ to $100,000$ iterations to obtain 6 significant digit of accuracy.
- CG will produce the exact solution in the absence of round off error in $\simeq 10,000$ iterations, however,
- Since $\sqrt{cond(A)} \simeq 100$, CG will produce an approximate solution with 6 significant digits of accuracy in $\simeq 500 - 1000$ iterations!
- With simple preconditioners (a topic that is coming) we get 6 digits in $\simeq 30 - 40$ iterations!

Exercise 8.1. *Write a computer program implementing Algorithm 8.2. Write the program so it can be applied to any given matrix A, with any given initial guess x^0 and right side b. Assume the iteration is converged when both the conditions $\|r^n\| < \epsilon\|b\|$ and $\|u^n - u^{n-1}\| < \epsilon\|u^n\|$ are satisfied for given tolerance ϵ. Consider the matrix*

$$A_1 = \begin{bmatrix} 2 & -1 & 0 & 0 & 0 \\ -1 & 2 & -1 & 0 & 0 \\ 0 & -1 & 2 & -1 & 0 \\ 0 & 0 & -1 & 2 & -1 \\ 0 & 0 & 0 & -1 & 2 \end{bmatrix}.$$

(1) Apply your program to the matrix A_1 using the exact solution $x_{\text{exact}} = [1,2,3,4,5]^t$ and $b_1 = A_1 x_{\text{exact}}$, starting with $x^0 = x_{\text{exact}}$. Demonstrate convergence to x_{exact} in a single iteration with $\epsilon = 10^{-4}$.

---

[3] All previous d^n must be stored and used in order to compute x^{n+1}.

(2) *Apply your program to A_1 and b_1 with tolerance $\epsilon = 10^{-4}$ but with initial guess $x^0 = 0$. Demonstrate convergence to x_{exact} in no more than five iterations.*

(3) *Repeat the previous two cases with the matrix*

$$A_2 = \begin{bmatrix} 2 & -1 & 0 & -1 & 0 \\ -1 & 3 & -1 & 0 & -1 \\ 0 & -1 & 2 & -1 & 0 \\ -1 & 0 & -1 & 3 & -1 \\ 0 & -1 & 0 & -1 & 3 \end{bmatrix}.$$

Exercise 8.2. *Recall that Exercises 7.15 (page 169) and 7.16 (page 170) had you wrote programs implementing iterative methods for the 2D MPP.*

Write a computer program to solve the 2D MPP using conjugate gradients Algorithm 8.2. How many iterations does it take to converge when $N = 100$ and $\epsilon = 1.e - 8$?

Recommendation: *You have already written and tested a conjugate gradient code in Exercise 8.1 and a 2D MPP code in Exercise 7.16. If you replace the matrix-vector products Ax^n appearing in your conjugate gradient code with function or subroutine calls that use 2D MPP code to effectively compute the product without explicitly generating the matrix A, you can leverage your earlier work and save development and debugging time.*

8.1.1 *Algorithmic options*

There are many algorithmic options (we will list two below) but the above is a good, stable and efficient form of CG.

(1) An equivalent expression for α_n is

$$\alpha_n = \frac{\langle r^n, r^n \rangle}{\langle d^n, Ad^n \rangle}.$$

(2) The expression $r^{n+1} = b - Ax^{n+1}$ is equivalent to $r^{n+1} = r^n - \alpha_n Ad^n$ in exact arithmetic. To see it is equivalent, we note that the residuals satisfy their own iteration.

Lemma 8.1. *In exact arithmetic, the CG residuals satisfy*

$$r^{n+1} = r^n - \alpha_n Ad^n.$$

Proof. Since $x^{n+1} = x^n + \alpha_n d^n$, multiply by $-A$ and add b to both sides. This gives

$$\underbrace{b - Ax^{n+1}}_{r^{n+1}} = \underbrace{b - Ax^n}_{r^n} - \alpha_n A d^n,$$

as claimed above. $\qquad\square$

Thus, the residual can be calculated 2 ways: directly at the cost of an extra matrix-vector multiply and via the above step. Direct calculation doubles the number of matrix-vector multiplies per step over $r^{n+1} = r^n - \alpha_n A d^n$. Some have reported that for highly ill-conditioned systems, it can be preferable to calculate the residual directly, possibly even in extended precision.

Exercise 8.3. *Consider the CG method, Algorithm 8.2. Show that it can be written as a three term recursion of the general form $x^{n+1} = \alpha_n x^n + \beta_n x^{n-1} + c^n$.*

Exercise 8.4. *In Exercise 8.1, you wrote a computer program to implement Algorithm 8.2. Double-check it on the 5×5 SPD matrix A_1 from that exercise by choosing any vector b and verify that the system $Ax = b$ can be solved in five iterations.*

Make a copy of your program and modify it to include the alternative expressions for α_n and r^n described above. Verify that the modified program gives rise to the same sequence of coefficients α_n and β_n, iterates x^n and residuals r^n as the original.

8.1.2 *CG's two main convergence theorems*

"All sorts of computer errors are now turning up. You'd be surprised to know the number of doctors who claim they are treating pregnant men."

— Anonymous Official of the Quebec Health Insurance Board, on Use of Computers in Quebec Province's Comprehensive Medical-care system. F. 19, 4:5. In Barbara Bennett and Linda Amster, Who Said What (and When, and Where, and How) in 1971: December–June, 1971 (1972), Vol. 1, 38.

The global optimality properties of the CG method depends on a specific family of subspaces, the Krylov subspaces. First recall.

Definition 8.2. Let $z^1, \cdots, z^m$ be m vectors. Then $span\left\{z^1, \cdots, z^m\right\}$ is the set of all linear combinations of $z^1, \cdots, z^m$, i.e., the subspace

$$span\left\{z^1, \cdots, z^m\right\} = \left\{x = \sum_{i=1}^{m} \alpha_i z^i : \alpha_i \in \mathbb{R}\right\}.$$

It will be important to know the form of the CG iterates and search directions. To find the correct subspaces, we step through the algorithm:

$$d^0 = r^0$$
$$x^1 = x^0 + \alpha r^0 \in x^0 + span\{r^0\}.$$

From Lemma 8.1 we have

$$r^1 = r^0 - \alpha A r^0 \in r^0 + A \cdot span\{r^0\},$$

so

$$d^1 = r^1 + \beta r^0 = r^0 - \alpha A r^0 + \beta r^0 \in span\{r^0, A r^0\}.$$

Thus

$$x^2 = x^1 + \alpha d^1 = x^0 + \alpha r^0 + \tilde{\alpha}\{r^0 - \alpha A r^0 + \beta r^0\}, \text{ so}$$
$$x^2 \in x^0 + span\{r^0, A r^0\}, \text{ and similarly}$$
$$r^1 \in r^0 + A \cdot span\{r^0, A r^0\}.$$

Continuing, we easily find the following.

Proposition 8.1. *The CG iterates x^j, residuals r^j and search directions d^j satisfy*

$$x^j \in x^0 + span\{r^0, A r^0, \cdots, A^{j-1} r^0\},$$
$$r^j \in r^0 + A \cdot span\{r^0, A r^0, \cdots, A^{j-1} r^0\}$$
$$and$$
$$d^j \in span\{r^0, A r^0, \cdots, A^{j-1} r^0\}.$$

Proof. Induction. $\qquad\qquad\qquad\qquad\qquad\qquad\qquad\qquad\qquad\qquad\square$

The subspace and affine subspaces, known as Krylov subspaces, are critical to the understanding of the method.

Definition 8.3. Let x^0 be given and $r^0 = b - A x^0$. The Krylov subspace determined by r^0 and A is

$$X_n = X_n(A; r^0) = span\{r^0, A r^0, \ldots, A^{n-1} r^0\}$$

and the affine Krylov space determined by r^0 and A is

$$K_n = K_n(A; x^0) = x^0 + X_n = \{x^0 + x : x \in X_n\}.$$

The first important theorem of CG method is the following. It explains the global error minimization linked to the choice of search directions.

Theorem 8.1. *Let A be SPD. Then the CG method satisfies the following:*

(i) The nth residual is globally optimal over the affine subspace K_n in the A^{-1}-norm

$$||r^n||_{A^{-1}} = \min_{r \in r^0 + AX_n} ||r||_{A^{-1}}.$$

(ii) The nth error is globally optimal over K_n in the A-norm

$$||e^n||_A = \min_{e \in K_n} ||e||_A.$$

(iii) $J(x^n)$ is the global minimum over K_n

$$J(x^n) = \min_{x \in K_n} J(x).$$

(iv) Furthermore, the residuals are orthogonal and search directions are A-orthogonal:

$$r^k \cdot r^l = 0 \quad , \quad \text{for } k \neq l,$$
$$\langle d^k, d^l \rangle_A = 0 \quad , \quad \text{for } k \neq l.$$

These are algebraic properties of CG iteration, proven by induction. Part (iv) already implies the finite termination property.

Exercise 8.5. *Prove the theorem by induction, starting from Algorithm 8.2. You may find Lemma 8.1 helpful.*

Corollary 8.1. *Let A be SPD. Then in exact arithmetic CG produces the exact solution to an $N \times N$ system in N steps or fewer.*

Proof. Since the residuals $\{r^0, r^1, \dots, r^{N-1}\}$ are orthogonal they are linearly independent. Thus, $r^l = 0$ for some $l \leq N$. $\qquad\square$

Using the properties (i) through (iv), the error in the nth CG step will be linked to an analytic problem: the error in Chebychev interpolation. The main result of it is the second big convergence theorem for CG.

Theorem 8.2. *Let A be SPD. Given any $\varepsilon > 0$ for*

$$n \geq \frac{1}{2}\sqrt{\text{cond}(A)} \ln\left(\frac{2}{\varepsilon}\right) + 1$$

the error in the CG iterations is reduced by ε:

$$||x^n - x||_A \leq \varepsilon ||x^0 - x||_A.$$

8.2 Analysis of the CG Algorithm

> Art has a double face, of expression and illusion, just like science
> has a double face: the reality of error and the phantom of truth.
> — René Daumal
> 'The Lie of the Truth'. (1938) translated by Phil Powrie (1989).
> In Carol A. Dingle, Memorable Quotations (2000).

The form of the CG algorithm presented in the last section is quite
computationally efficient. It has developed over some years as many equivalences and identities have been derived for the method. We give two
different (but of course equivalent if you look deeply enough) analytical
developments of CG. The first is a straightforward application of the
Pythagorean theorem. CG sums an orthogonal series and the orthogonal
basis vectors are generated by a special method, the Orthogonalization of
Moments algorithm. Putting these two together immediately gives a simplified CG method which has the essential and remarkable features claimed
for it.

The second approach is indirect and more geometric. In this second
approach, we shall instead define the CG method by (CG as n dimensional
minimization). This definition makes the global optimization property obvious. However, it also suggests that the nth step requires an n dimensional
optimization calculation. Thus the work in this approach will be to show
that the n dimensional optimization problem can be done by a 1 dimensional line search. In other words, it will be to show that the n dimensional
optimization problem can be done by either one 3 term recursion or two
coupled 2-term recursions. This proves that (CG as n dimensional minimization) can be written in the general form announced in the introduction.
The key to this second approach is again the Orthogonalization of Moments
algorithm.

Since any treatment will adopt one or the other, the Orthogonalization
of Moments algorithm will be presented twice.

8.3 Convergence by the Projection Theorem

> Fourier is a mathematical poem.
> — Thomson, [Lord Kelvin] William (1824–1907)

We begin with some preliminaries. The best approximation under a
norm given by an inner product in a subspace is exactly the orthogonal

projection with respect to that inner product. Thus we start by recalling some fundamental properties of these best approximations. Let $X \subset \mathbb{R}^N$ be an n (for $n < N$) dimensional subspace and $x \in \mathbb{R}^N$. Given an inner product and associated norm[4] $\langle \cdot, \cdot \rangle_*$, $\| \cdot \|_* = \langle \cdot, \cdot \rangle_*^{1/2}$, the **best approximation** $x^n \in X$ to x, is the unique $x^n \in X$ satisfying:

$$\|x - x^n\|_* = \min_{\widetilde{x} \in X} \|x - \widetilde{x}\|_*.$$

If K is the affine space $K = x^0 + X$ (where x^0 is fixed), then the best approximation in K is the solution to

$$\|x - x^K\|_* = \min_{\widetilde{x} \in K} \|x - \widetilde{x}\|_*.$$

The two best approximations are related. Given x, x^K, the best approximation in K, is given by $x^K = x^0 + x^X$ where x^X is the best approximation in X to $x - x^0$.

Theorem 8.3 (Pythagorean or Projection Theorem). *Let X be a subspace of $\mathbb{R}^n$ and $x \in \mathbb{R}^N$. Then, the best approximation to x in X, $x^n \in X$*

$$\|x - x^n\|_* = \min_{\widetilde{x} \in X} \|x - \widetilde{x}\|_*$$

is determined by

$$\langle x - x^n, \widetilde{x} \rangle_* = 0, \qquad \forall \widetilde{x} \in X.$$

Further, we have

$$\|x\|_*^2 = \|x^n\|_*^2 + \|x - x^n\|_*^2.$$

Let $x^0 \in \mathbb{R}^N$ and let K be an affine sub-space $K = x^0 + X$. Given $x \in \mathbb{R}^N$ there exists a unique best approximation $x^n \in K$ to x:

$$\|x - x^n\|_* = \min_{\widetilde{x} \in K} \|x - \widetilde{x}\|_*.$$

The error is orthogonal to X

$$\langle x - x^n, \widetilde{x} \rangle_* = 0, \qquad \forall \widetilde{x} \in X.$$

Proof. See any book on linear algebra! □

---

[4]The innerproduct is "fixed but arbitrary". Think of the usual dot product and the A-inner product for concrete examples.

The best approximation in $K = x^0 + X$ is determined by $\langle x - x^n, \widetilde{x} \rangle_* = 0, \forall \widetilde{x} \in X$. Let $e_1, \ldots, e_n$ be a basis for X. Expand $x^n = x^0 + \sum c_j e^j \in K$ then the vector of undetermined coefficients satisfies the linear system

$$Gc = f, \qquad G_{ij} = \langle e^i, e^j \rangle_*, \qquad f_j = \langle x - x^0, e^j \rangle_*.$$

Here G is called the "**Gram matrix**" or "Gramian".

Definition 8.4. Let $\{\phi^1, \phi^2, \ldots, \phi^n\}$ be a basis for X and $\langle \cdot, \cdot \rangle_*$ an inner product on X. The associated **Gram matrix** G of the basis is

$$G_{ij} = \langle \phi^i, \phi^j \rangle_*.$$

Thus, the general way to calculate the best approximation in an n-dimensional affine subspace $K = x^0 + X$ is to pick a basis for X, assemble the $n \times n$ Gram matrix and solve an $n \times n$ linear system. Two questions naturally arise:

- How to calculate all the inner products f_j if x is the sought but unknown solution of $Ax = b$?
- Are there cases when the best approximation in K can be computed at less cost than constructing a basis, assembling G and then solving $Gc = f$?

For the first question there is a clever finesse that works when A is SPD. Indeed, if we pick $\langle \cdot, \cdot \rangle_* = \langle \cdot, \cdot \rangle_A$ then for $Ax = b$,

$$\langle x, y \rangle_A = x^t A y = (Ax)^t y = b^t y = \langle b, y \rangle$$

which is computable without knowing x. For the second question, there is a case when calculating the best approximation is easy: *when an orthogonal basis is known for X*. This case is central to many mathematical algorithms including CG.

Definition 8.5 (Orthogonal basis). Let $\{\phi^1, \phi^2, \ldots, \phi^n\}$ be a basis for X and $\langle \cdot, \cdot \rangle_*$ an inner produce on X. An **orthogonal basis** for X is a basis $\{\phi^1, \phi^2, \ldots, \phi^n\}$ satisfying additionally

$$\langle \phi^i, \phi^j \rangle_* = 0 \text{ whenever } i \neq j.$$

If the basis $\{\phi^1, \phi^2, \ldots, \phi^n\}$ is orthogonal then its Gram matrix G_{ij} is diagonal. The best approximation in X can then simply be written down explicitly

$$x^{n+1} = \sum_{j=1}^{n} \frac{\langle x, \phi_j \rangle_*}{\langle \phi_j, \phi_j \rangle_*} \phi_j.$$

Similarly, the best approximation in the affine subspace K can also be written down explicitly as

$$x^{n+1} = y^0 + \sum_{j=1}^{n} \frac{\langle x - y^0, \phi_j \rangle_*}{\langle \phi_j, \phi_j \rangle_*} \phi_j.$$

Summing this series for the best approximation in an affine subspace can be expressed as an algorithm that looks like (and is) a descent method.

Algorithm 8.3 (Summing an orthogonal series). *Given $x \in \mathbb{R}^N$, an n-dimensional subspace X, with orthogonal basis $\{\phi_1, \cdots, \phi_n\}$ and a vector x^0*

$x^0 = y^0$
for j=0:n-1
 $d^j = \phi^{j+1}$
 $\alpha_j = \langle x - x^0, d^j \rangle_* / \langle d^j, d^j \rangle_*$
 $x^{j+1} = x^j + \alpha_j d^j$
end

The usual descent method (general directions) produces at each step an approximation optimal in the (1-dimensional) line $x = x^j + \alpha d^j, \alpha \in \mathbb{R}$. Since the descent directions are orthogonal this produces at the jth step an approximation that is optimal over the j-dimensional affine subspace $x^0 + span\{\phi_1, \ldots, \phi_j\}$.

Theorem 8.4. *If $\{\phi_1, \ldots, \phi_j\}$ are A-orthogonal and $\langle \cdot, \cdot \rangle_*$ is the A-inner product, then the approximations produced by the descent method choosing $\{\phi_1, \ldots, \phi_j\}$ for descent directions (i.e., if choosing $d^i = \phi^i$) are the same as those produced by summing the orthogonal series in Algorithm 8.3 above. Thus, with A-orthogonal search directions, the approximations produced by the descent algorithm satisfy*

$$\|x - x^j\|_A = \min_{\tilde{x} \in x^0 + span\{\phi_1, \ldots, \phi_j\}} \|x - \tilde{x}\|_A,$$

$$J(x^j) = \arg \min_{\tilde{x} \in x^0 + span\{\phi_1, \ldots, \phi_j\}} J(\tilde{x}).$$

Proof. Thus, consider the claim of equivalence of the two methods. The general step of each takes the form $x^{j+1} = x^j + \alpha d^j$ with the same x^j, d^j. We thus need to show equivalence of the two formulas for the stepsize:

$$\text{descent}: \alpha_j = \langle r^j, \phi^j \rangle / \langle \phi^j, \phi^j \rangle_A$$

$$\text{orthogonal series}: \quad \alpha_j = \langle x - y^0, \phi^j \rangle_A / \langle \phi^j, \phi^j \rangle_A.$$

Since the denominators are the same we begin with the first numerator and show its equal to the second. Indeed,

$$\langle r^j, \phi^j \rangle = \langle b - Ax^j, \phi^j \rangle = \langle Ax - Ax^j, \phi^j \rangle$$
$$= \langle x - x^j, \phi^j \rangle_A.$$

Consider the form of x^j produced by the descent algorithm. We have (both obvious and easily proven by induction) that x^j takes the general form

$$x^j = x^0 + a_1 \phi^1 + \cdots + a_{j-1} \phi^{j-1}.$$

Thus, by A-orthogonality of $\{\phi_1, \ldots, \phi_j\}$

$$\langle x^j, \phi^j \rangle_A = \langle x^0 + a_1 \phi^1 + \cdots + a_{j-1} \phi^{j-1}, \phi^j \rangle_A = \langle x^0, \phi^j \rangle_A.$$

Thus we have

$$\langle r^j, \phi^j \rangle = \langle x - x^j, \phi^j \rangle_A = \langle x - x^0, \phi^j \rangle_A,$$

which proves equivalence. The error estimate is just restating the error estimate of the Pythagorean theorem. From the work on descent methods we know that A-norm optimality of the error is equivalent to minimization of $J(\cdot)$ over the same space. Hence the last claim follows. □

Thus:

- Algorithm 8.3 does 1-dimensional work at each jth step (a 1-dimensional optimization) and attains a j-dimensional optimum error level;
- Equivalently, if the descent directions are chosen A-orthogonal, a j-dimensional minimizer results.

The focus now shifts to how to generate the orthogonal basis. The classic method is the Gram–Schmidt algorithm.

8.3.1 The Gram–Schmidt algorithm

The Gram–Schmidt algorithm is not used in CG for SPD systems. It is important for understanding the method actually used (orthogonalization of moments which is coming) and becomes important in generalized conjugate gradient methods for nonsymmetric systems. For example, GS is used to generate search directions in the method GMres.

Algorithm 8.4 (Gram–Schmidt orthogonalization). *Given a basis* $\{e^1, e^2, \ldots, e^N\}$ *for* $\mathbb{R}^N$,

```
φ¹ = e¹
for j=1:n
  for i=1:j
    αᵢ = ⟨eʲ⁺¹, φⁱ⟩* / ⟨φⁱ, φⁱ⟩*
  end
  φʲ⁺¹ = eʲ⁺¹ - Σⱼ₌₁ⁱ αᵢφⁱ
end
```

Theorem 8.5. *Given a basis $\{e^1, e^2, \ldots, e^N\}$ for $\mathbb{R}^N$, the Gram–Schmidt Algorithm 8.4 constructs a new, $\langle \cdot, \cdot \rangle_*$-orthogonal basis $\phi_1, \ldots, \phi_N$ for $\mathbb{R}^N$ so that:*

(1) $span\{e^1, \ldots, e^j\} = span\{\phi_1, \ldots, \phi_j\}$ for each $j = 1, \cdots, N$; and,
(2) $\langle \phi_i, \phi_j \rangle_* = 0$ whenever $i \neq j$.

The nth step of Gram–Schmidt obtains an orthogonal basis for an n-dimensional subspace as a result of doing n-dimensional work calculating the n coefficients $\alpha_i, i = 1, \cdots, n$. There is exactly one case where this work can be reduced dramatically and that is the case relevant for the conjugate gradient method. Since summing an orthogonal series is globally optimal but has 1-dimensional work at each step, the problem shifts to finding an algorithm for constructing an A-orthogonal basis which, unlike Gram–Schmidt, requires 1-dimensional work at each step. There is exactly one such method which only works in exactly one special case (for the Krylov subspace of powers of A times a fixed vector) called *"Orthogonalization of moments"*.

Exercise 8.6. *Prove that the Gram matrix $G_{ij} = \langle e^i, e^j \rangle_*$ is SPD provided $e^1, \ldots, e^n$ is a basis for X and diagonal provided $e^1, \ldots, e^n$ are orthogonal.*

Exercise 8.7. *Give an induction proof that the Gram–Schmidt Algorithm 8.4 constructs a new basis with the two properties claimed in Theorem 8.5.*

8.3.2   *Orthogonalization of moments instead of Gram–Schmidt*

A great part of its [higher arithmetic] theories derives an additional charm from the peculiarity that important propositions, with the impress of simplicity on them, are often easily discovered by induction, and yet are of so profound a character that we cannot find the demonstrations till after many vain attempts; and even then, when we do succeed, it is often by some tedious and artificial process, while the simple methods may long remain concealed.

— Gauss, Karl Friedrich (1777–1855).

In H. Eves Mathematical Circles Adieu, Boston: Prindle, Weber and Schmidt, 1977.

The CG method at the *nth* step computes an A-norm optimal approximation in a n-dimensional subspace. In general this requires solving an $N \times N$ linear system with the Gram matrix. The only case, and the case of the CG method, when it can be done with much less expense is when an A-orthogonal basis is known for the subspace and this is known only for a Krylov subspace:

$$X_n := span\{r^0, Ar^0, A^2r^0, \cdots, A^{n-1}r^0\}$$
$$K_n := x^0 + X_n.$$

CG hinges on **an efficient method of determining an A-orthogonal basis for X_n.** With such a method, CG takes the general form:

Algorithm 8.5. *Given SPD A, initial guess x^0, and maximum number of iterations* `itmax`,

$r^0 = b - Ax^0$
$d^0 = r^0$
for n=1:itmax
   Descent step:
     $\alpha_{n-1} = \langle r^{n-1}, d^{n-1}\rangle / \langle d^{n-1}, d^{n-1}\rangle_A$
     $x^n = x^{n-1} + \alpha_{n-1}d^{n-1}$
     $r^n = b - Ax^n$
   OM step:
     Calculate new A-orthogonal search direction d^n so that
     $span\{d^0, d^1, \ldots, d^n\} = span\{r^0, Ar^0, A^2r^0, \ldots, A^nr^0\}$
end

The key (OM step) is accomplished by the "Orthogonalization of moments" algorithm, so-called because *moments of an operator A* are powers of A acting on a fixed vector. This algorithm takes a set of moments $\{e^1, e^2, e^3, \cdots, e^j\}$ where $e^j = A^{j-1}e^1$ and generates an A-orthogonal basis $\{\phi^1, \phi^2, \phi^3, \cdots, \phi^j\}$ spanning the same subspace.

Algorithm 8.6 (Orthogonalization of moments algorithm). *Let* A *be SPD, and* $e^1 \in \mathbb{R}^N$ *be a given vector.*

$\phi^1 = e^1$
for n=1:N-1
    $\alpha = \langle \phi^n, A\phi^n \rangle_A / \langle \phi^n, \phi^n \rangle_A$
    if n==1
        $\phi^2 = A\phi^1 - \alpha\phi^1$
    else
        $\beta = \langle \phi^{n-1}, A\phi^n \rangle_A / \langle \phi^{n-1}, \phi^{n-1} \rangle_A$
        $\phi^{n+1} = A\phi^n - \alpha\phi^n - \beta\phi^{n-1}$
    end
end

In comparison with Gram–Schmidt, this algorithm produces an A-orthogonal basis of the Krylov subspace through a three term recursion.

Theorem 8.6 (Orthogonalization of moments). *Let* A *be an SPD matrix. The sequence* $\{\phi^j\}_{j=1}^N$ *produced by the Orthogonalization of Moments Algorithm 8.6 is* A*-orthogonal. Further, for* $e^j = A^{j-1}e^1$, *at each step* $1 \le j \le N$

$$span\{e^1, e^2, e^3, \cdots, e^j\} = span\{\phi^1, \phi^2, \phi^3, \cdots, \phi^j\}.$$

Proof. Preliminary remarks: First note that the equation for ϕ^{n+1} takes the form

$$\phi^{n+1} = A\phi^n + \alpha\phi^n + \beta\phi^{n-1}. \tag{8.1}$$

Consider the RHS of this equation. We have, by the induction hypothesis

$$A\phi^n \in span\{e^1, e^2, e^3, \cdots, e^{n+1}\}$$
$$\alpha\phi^n + \beta\phi^{n-1} \in span\{e^1, e^2, e^3, \cdots, e^n\},$$
$$\text{and } \langle \phi^n, \phi^{n-1} \rangle_A = 0.$$

The step $\phi^{n+1} = A\phi^n + \alpha\phi^n + \beta\phi^{n-1}$ contains two parameters. It is easy to check that the parameters α and β are picked (respectively) to make the two equations hold:

$$\langle \phi^{n+1}, \phi^n \rangle_A = 0,$$
$$\langle \phi^{n+1}, \phi^{n-1} \rangle_A = 0.$$

Indeed

$$0 = \left\langle \phi^{n+1}, \phi^n \right\rangle_A = \left\langle A\phi^n + \alpha\phi^n + \beta\phi^{n-1}, \phi^n \right\rangle_A$$
$$= \left\langle A\phi^n, \phi^n \right\rangle_A + \alpha \left\langle \phi^n, \phi^n \right\rangle_A + \beta \left\langle \phi^{n-1}, \phi^n \right\rangle_A$$
$$= \left\langle A\phi^n, \phi^n \right\rangle_A + \alpha \left\langle \phi^n, \phi^n \right\rangle_A$$

and the same for $\langle \phi^{n+1}, \phi^{n-1} \rangle_A = 0$ gives two equations for α, β whose solutions are exactly the values chosen on Orthogonalization of Moments. The key issue in the proof is thus to show that

$$\langle \phi^{n+1}, \phi^j \rangle_A = 0, \text{ for } j = 1, 2, \cdots, n-2. \tag{8.2}$$

This will hold precisely because $span\{e^1, e^2, e^3, \cdots, e^j\}$ is a Krylov subspace determined by moments of A.

Details of the proof: We show from (8.1) that (8.2) holds. The proof is by induction. To begin, from the choice of α, β it follows that the theorem holds for $j = 1, 2, 3$. Now suppose the theorem holds for $j = 1, 2, \cdots, n$. From (8.1) consider $\langle \phi^{n+1}, \phi^j \rangle_A$. By the above preliminary remarks, this is zero for $j = n, n-1$. Thus consider $j \leq n-2$. We have

$$\langle \phi^{n+1}, \phi^j \rangle_A = \langle A\phi^n, \phi^j \rangle_A + \alpha\langle \phi^n, \phi^j \rangle_A + \beta\langle \phi^{n-1}, \phi^j \rangle_A$$
$$\text{for } j = 1, 2, \cdots, n-2.$$

By the induction hypothesis

$$\langle \phi^n, \phi^j \rangle_A = \langle \phi^{n-1}, \phi^j \rangle_A = 0$$

thus it simplifies to

$$\langle \phi^{n+1}, \phi^j \rangle_A = \langle A\phi^n, \phi^j \rangle_A.$$

Consider thus $\langle A\phi^n, \phi^j \rangle_A$. Note that A is self adjoint with respect to the A-inner product. Indeed, we calculate

$$\langle A\phi^n, \phi^j \rangle_A = (A\phi^n)^t A\phi^j = (\phi^n)^t A^t A\phi^j = (\phi^n)^t AA\phi^j = \langle \phi^n, A\phi^j \rangle_A.$$

Thus, $\langle A\phi^n, \phi^j \rangle_A = \langle \phi^n, A\phi^j \rangle_A$. By the induction hypothesis (and because we are dealing with a Krylov subspace): for $j \leq n-2$

$$\phi^j \in span\{e^1, e^2, e^3, \cdots, e^{n-2}\}$$

thus

$$A\phi^j \in span\{e^1, e^2, e^3, \cdots, e^{n-2}, e^{n-1}\}.$$

Further, by the induction hypothesis

$$span\{e^1, e^2, e^3, \cdots, e^{n-1}\} = span\{\phi^1, \phi^2, \phi^3, \cdots, \phi^{n-1}\}.$$

Finally by the induction hypothesis $\{\phi^1, \phi^2, \phi^3, \cdots, \phi^n\}$ are A-orthogonal, so

$$\langle \phi^n, \text{ something in } span\{\phi^1, \phi^2, \phi^3, \cdots, \phi^{n-1}\}\rangle_A = 0.$$

Putting the steps together:

$$\langle \phi^{n+1}, \phi^j\rangle_A = \langle A\phi^n, \phi^j\rangle_A = \langle \phi^n, A\phi^j\rangle_A$$
$$= \langle \phi^n, \text{ something in } span\{\phi^1, \phi^2, \phi^3, \cdots, \phi^{n-1}\}\rangle_A = 0.$$

All that remains is to show that

$$span\{e^1, e^2, e^3, \cdots, e^{n+1}\} = span\{\phi^1, \phi^2, \phi^3, \cdots, \phi^{n+1}\}.$$

This reduces to showing $e^{n+1} = Ae^n \in span\{\phi^1, \phi^2, \phi^3, \cdots, \phi^{n+1}\}$. It follows from the induction hypothesis and (8.1) and is left as an exercise. $\qquad\square$

The orthogonalization of moments algorithm is remarkable. Using it to find the basis vectors [which become the descent directions] and exploiting the various orthogonality relations, we shall see that the CG method simplifies to a very efficient form.

Exercise 8.8. (An exercise in looking for similarities in different algorithms). *Compare the orthogonalization of moments algorithm to the one (from any comprehensive numerical analysis book) which produces orthogonal polynomials. Explain their similarities.*

Exercise 8.9. *If A is not symmetric, where does the proof break down? If A is not positive definite, where does it break down?*

Exercise 8.10. *Write down the Gram–Schmidt algorithm for producing an A-orthogonal basis. Calculate the complexity of Gram–Schmidt and Orthogonalization of moments. (Hint: Count matrix-vector multiplies and inner products, ignore other operations.) Compare.*

Exercise 8.11. *Complete the proof of the Orthogonalization of Moments Theorem.*

8.3.3 *A simplified CG method*

We can already present a CG method that attains the amazing properties claimed of CG in the first section. The method is improvable in various ways, but let us first focus on the major advances made by just descent (equivalent to summing an orthogonal series) with the directions generated by orthogonalization of moments. Putting the two together in a very simple way gives the following version of the CG method.

Algorithm 8.7 (A version of CG). *Given SPD A and initial guess x^0,*

$r^0 = b - Ax^0$

$d^0 = r^0/\|r^0\|$

First descent step:

$$\alpha_0 = \frac{\langle d^0, r^0 \rangle}{\langle d^0, d^0 \rangle_A}$$

$x^1 = x^0 + \alpha_0 d^0$

$r^1 = b - Ax^1$

First step of OM:

$$\gamma_0 = \frac{\langle d^0, Ad^0 \rangle_A}{\langle d^0, d^0 \rangle_A}$$

$d^1 = Ad^0 - \gamma_0 d^0$

$d^1 = d^1/\|d^1\|$ *(normalize[5] d^1)*

for n=1:∞

  Descent Step:

$$\alpha_n = \frac{\langle r^n, d^n \rangle}{\langle d^n, d^n \rangle_A}$$

$x^{n+1} = x^n + \alpha_n d^n$

$r^{n+1} = b - Ax^{n+1}$

if *converged, STOP,* end

OM step:

$$\gamma_n = \frac{\langle d^n, Ad^n \rangle_A}{\langle d^n, d^n \rangle_A}$$

$$\beta_n = \frac{\langle d^{n-1}, Ad^n \rangle_A}{\langle d^{n-1}, d^{n-1} \rangle_A}$$

$d^{n+1} = Ad^n - \gamma_n d^n - \beta_n d^{n-1}$

$d^{n+1} = d^{n+1}/\|d^{n+1}\|$ *(normalize[5] d^{n+1})*

end

Algorithm 8.7, while not the most efficient form for computations, captures the essential features of the method. The differences between the above version and the highly polished one given in the first section, Algorithm 8.2, take advantage of the various orthogonality properties of CG. These issues, while important, will be omitted to move on to the error analysis of the method.

---

[5]Normalizing d is not theoretically necessary, but on a computer, d could grow large enough to cause the calculation to fail.

Exercise 8.12. *Consider the above version of CG. Show that it can be written as a three term recursion of the general form $x^{n+1} = a_n x^n + b_n x^{n-1} + c_n$.*

Exercise 8.13. *In Exercise 8.2, you wrote a program implementing CG for the 2D MPP and compared it with other iterative methods. Make a copy of that program and modify it to implement the simplified version of CG given in Algorithm 8.7. Show by example that the two implementations are equivalent in the sense that they generate essentially the same sequence of iterates.*

8.4   Error Analysis of CG

> But as no two (theoreticians) agree on this (skin friction) or any other subject, some not agreeing today with what they wrote a year ago, I think we might put down all their results, add them together, and then divide by the number of mathematicians, and thus find the average coefficient of error.
> — Sir Hiram Maxim, In Artificial and Natural Flight (1908), 3.   Quoted in John David Anderson, Jr., Hypersonic and High Temperature Gas Dynamics (2000), 335.

> "To err is human but it feels divine."
> — Mae West

We shall show that the CG method takes $O(\sqrt{cond(A)})$ steps per significant digit (and, as noted above, at most N steps). This result is built up in several steps that give useful detail on error behavior. The first step is to relate the error to a problem in Chebychev polynomial approximation. Recall that we denote the polynomials of degree $\leq n$ by

$$\Pi_n := \{p(x) : p(x) \text{ is a real polynomial of degree } \leq n\}.$$

Theorem 8.7. *Let A be SPD. Then the CG method's error $e^n = x - x^n$ satisfies:*
(i) $e^n \in e^0 + span\{Ae^0, A^2 e^0, \cdots, A^n e^0\}$.
(ii) $||e^n||_A = \min\{||e||_A : e \in e^0 + span\{Ae^0, A^2 e^0, \cdots, A^n e^0\}\}$.
(iii) The error is bounded by

$$||x - x^n||_A \leq \left(\min_{p_n \in \Pi_n \ and \ p(0)=1} \max_{\lambda_{\min} \leq x \leq \lambda_{\max}} |p(x)| \right) ||e^0||_A.$$

Proof. As $Ae = r$ and

$$r^n \in r^0 + span\{Ar^0, A^2r^0, A^3r^0, \cdots, A^nr^0\}$$

this implies

$$Ae^n \in A\left(e^0 + span\{Ae^0, A^2e^0, A^3e^0, \cdots, A^ne^0\}\right),$$

which proves part (i). For (ii) note that since $||e^n||_A^2 = 2(J(x^n) - J(x))$ minimizing $J(x)$ is equivalent to minimizing the A-norm of e. Thus, part (ii) follows. For part (iii), note that part (i) implies

$$e^n = [I + a_1A + a_2A^2 + \cdots + a_nA^n]e^0 = p(A)e^0,$$

where $p(x)$ is a real polynomial of degree $\leq n$ and $p(0) = 1$. Thus, from this observation and part (ii),

$$||x - x^n||_A = \min_{p_n \in \Pi_n \text{ and } p(0)=1} ||p(A)e^0||_A$$

$$\leq \left(\min_{p_n \in \Pi_n \text{ and } p(0)=1} ||p(A)||_A\right) ||e^0||_A.$$

The result follows by calculating using the spectral mapping theorem that

$$||p(A)||_A = \max_{\lambda \in spectrum(A)} |p(x)| \leq \max_{\lambda_{\min} \leq x \leq \lambda_{\max}} |p(x)|.$$

$\square$

To determine the rate of convergence of the CG method, the question now is:

How big is the quantity:

$$\min_{\substack{p_n \in \Pi_n \\ p(0) = 1}} \max_{\lambda \in spectrum(A)} |p(\lambda)| \, ?$$

Fortunately, this is a famous problem of approximation theory, solved long ago by Chebychev.

The idea of Chebychev's solution is to pick points x_j in the interval $\lambda_{\min} \leq x \leq \lambda_{\max}$ and let $\widetilde{p}_n(x)$ interpolate zero at those points; $\widetilde{p}_n(x)$ solves the interpolation problem:

$$\widetilde{p}_n(0) = 1$$

$$\widetilde{p}_n(x_j) = 0, \quad j = 1, 2, \ldots, n, \text{ where } \lambda_{\min} \leq x_j \leq \lambda_{\max}.$$

By making $\widetilde{p}_n(x)$ zero at many points on $\lambda_{\min} \leq x \leq \lambda_{\max}$ we therefore force $\widetilde{p}_n(x)$ to be small over all of $\lambda_{\min} \leq x \leq \lambda_{\max}$. We have then

$$\min_{\substack{p_n \in \Pi_n \\ p(0) = 1}} \max_{\lambda \in spectrum(A)} |p(\lambda)| \leq \min_{\substack{p_n \in \Pi_n \\ p(0) = 1}} \max_{\lambda_{\min} \leq x \leq \lambda_{\max}} |p(x)|$$

$$\leq \max_{\lambda_{\min} \leq x \leq \lambda_{\max}} |\widetilde{p}_n(x)|.$$

The problem now shifts to finding the "best" points to interpolate zero, best being in the sense of the min-max approximation error. This problem is a classical problem of approximation theory and was also solved by Chebychev, and the resulting polynomials are called "Chebychev polynomials", one of which is depicted in Figure 8.1.

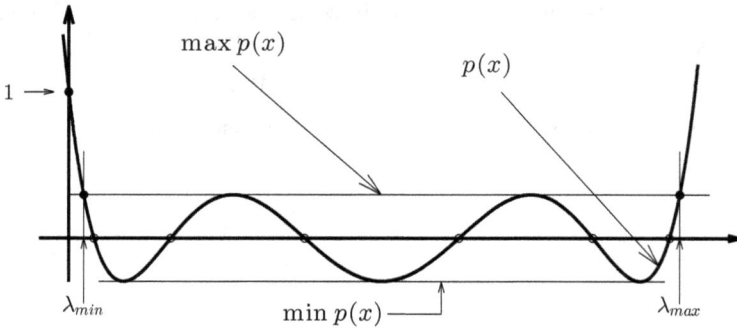

Fig. 8.1 We make $p(x)$ small by interpolating zero at points on the interval. In this illustration, the minimum and maximum values of $p(x)$ are computed on the interval $[\lambda_{min}, \lambda_{max}]$.

Theorem 8.8 (Chebychev polynomials, min-max problem). *The points x_j for which $\widetilde{p}_n(x)$ attains*

$$\min_{\substack{p_n \in \Pi_n \\ p(0)=1}} \max_{\lambda_{\min} \leq x \leq \lambda_{\max}} |p(x)|$$

are the zeroes on the Chebychev polynomial

$$T_n\left(\frac{b+a-2x}{b-a}\right) \bigg/ T_n\left(\frac{b+a}{b-a}\right)$$

on $[a,b]$ (where $a = \lambda_{\min}, b = \lambda_{\max}$). We then have

$$\min_{\substack{p_n \in \Pi_n \\ p(0)=1}} \max_{a \leq x \leq b} |p(x)| = \max_{a \leq x \leq b} \left|\frac{T_n(\frac{b+a-2x}{b-a})}{T_n(\frac{b+a}{b-a})}\right| =$$

$$= 2\frac{\sigma^n}{1+\sigma^n}, \sigma := \frac{1-\sqrt{\frac{a}{b}}}{1+\sqrt{\frac{a}{b}}}$$

Proof. For the proof and development of the beautiful theory of Chebychev approximation see any general approximation theory book. □

To convert this general result to a prediction about the rate of convergence of CG, simply note that

$$\frac{a}{b} = \frac{\lambda_{\min}}{\lambda_{\max}} = cond(A).$$

Thus we have the error estimate for CG:

Theorem 8.9. *Let A be SPD. Then*

(1) The nth CG residual is optimal over K_n in the A^{-1} norm:

$$||r^n||_{A^{-1}} = \min_{r \in K_n} ||r||_{A^{-1}}$$

(2) The nth CG error is optimal over K_n in the A norm:

$$||e^n||_A = \min_{e \in K_n} ||e||_A;$$

(3) The nth CG energy functional is optimal over K_n:

$$J(x^n) = \min_{x \in K_n} J(x);$$

(4) We have the orthogonality relations

$$\langle r^k, r^j \rangle = 0, \quad k \neq j,$$
$$\langle d^k, d^j \rangle_A = 0, \quad k \neq j;$$

(5) Given any $\varepsilon > 0$ for

$$n \geq \frac{1}{2}\sqrt{cond(A)} \ln(\frac{2}{\varepsilon}) + 1$$

the error in the CG iterations is reduced by ε:

$$||x^n - x||_A \leq \varepsilon ||x^0 - x||_A.$$

Exercise 8.14. Construct an interpolating polynomial of degree $\leq N$ with $p(0) = 1$ and $p(\lambda_j) = 0, 1 \leq j \leq N$. Use this to give a second proof that CG gives the exact answer in N steps or less.

Exercise 8.15. Show that if A has $M < N$ distinct eigenvalues then the CG method converges to the exact solution in at most $M(< N)$ steps. Recall that

$$||\hat{x} - x^n||_A = \left(\min_{\substack{p_n \in \Pi_n \\ p(0) = 1}} \max_{\lambda \in spectrum(A)} |p(x)| \right) ||e^0||_A.$$

Then construct a polynomial $p(x)$ of degree M with $p(\lambda) = 0$ for all $\lambda \in spectrum(A)$.

8.5   Preconditioning

"Oh, what a difficult task it was.
To drag out of the marsh
the hippopotamus"
— Korney Ivanovic' Chukovsky

The idea of preconditioning is to "preprocess" the linear system to reduce the condition number of A. We pick an SPD matrix M, *for which it is very easy to solve $My = f$*, and replace

$$Ax = b \iff M^{-1}Ax = M^{-1}b.$$

PCG for $Ax = b$ is CG for $M^{-1}Ax = M^{-1}b$.[6] Of course we never invert M explicitly; every time M^{-1} is written it means "solve a linear system with coefficient matrix M". A few simple examples of preconditioners;

- $M = $ a diagonal matrix with $M_{ii} = \sum_{j=1}^{N} |a_{ij}|$,
- $M = $ the tridiagonal part of A

$$M_{ij} = a_{ij} \text{ for } j = i, i - 1, i + 1, \text{ and}$$
$$M_{ij} = 0 \text{ otherwise.}$$

- If $A = A_0 + B$ where A_0 is simpler than A and easy to invert, then pick $M = A_0$. Instead of picking A_0, we can also pick B: the entries in A to discard to get the preconditioner M.
- $M = \widetilde{L}\widetilde{L}^t$, a simple and cheap approximation to the LL^t decomposition of A.
- Any iterative method indirectly determines a preconditioner. Indeed, since M approximates A the solution of $M\rho = r$ approximates the solution of $A\rho = r$. If some other iterative method is available as a subroutine then an approximate solution to $A\rho = r$ can be calculated by doing a few steps of some (other) iterative method for the equation $A\rho = r$. This determines a matrix M (indirectly of course).

If we are given an effective preconditioner M, PCG can be simplified to the following attractive form.

---

[6]To be very precise, A SPD and M SPD does not imply $M^{-1}A$ is SPD. However, $M^{-1}A$ is similar to the SPD matrix $M^{-1/2}AM^{-1/2}$. We think of PCG for $Ax = b$ as CG for $M^{-1}Ax = M^{-1}b$. Again, to be very picky, in actual fact it is CG for the system $\left(M^{-1/2}AM^{-1/2}\right)y = M^{-1/2}b$ with SPD coefficient matrix $M^{-1/2}AM^{-1/2}$. The first step is, after writing down CG for this system to reverse the change of variable everywhere $y \Leftarrow M^{1/2}x$ and eliminate all the $M^{\pm 1/2}$.

Algorithm 8.8 (PCG Algorithm for solving $Ax = b$**).** *Given a SPD matrix A, preconditioner M, initial guess vector x^0, right side vector b, and maximum number of iterations* `itmax`

$$r^0 = b - Ax^0$$
$$Solve\ Md^0 = r^0$$
$$z^0 = d^0$$
`for n=0:itmax`
$$\alpha_n = \langle r^n, z^n\rangle/\langle d^n, Ad^n\rangle$$
$$x^{n+1} = x^n + \alpha_n d^n$$
$$r^{n+1} = b - Ax^{n+1} \quad (*)$$
`if` *converged, stop* `end`
$$Solve\ Mz^{n+1} = r^{n+1}$$
$$\beta_{n+1} = \langle r^{n+1}, z^{n+1}\rangle/\langle r^n, z^n\rangle$$
$$d^{n+1} = z^{n+1} + \beta_{n+1}d^n \quad (**)$$
`end`

Note that the extra cost is exactly one solve with M each step. There is a good deal of art in picking preconditioners that are inexpensive to apply and that reduce $cond(A)$ significantly.

Exercise 8.16. *Let $Ax = b$ be converted to a fixed point problem $(I - T)x = f$ (associated with a simple iterative method). If $I - T$ is SPD we can apply CG to this equation resulting in using a simple iterative method to precondition A. (a) Suppose $||T|| < 1$. Estimate $cond(I-T)$ in terms of $1 - ||T||$. (b) For the MPP pick a simple iterative method and work out for that method (i) if $I - T$ is SPD, and (ii) whether $cond(I - T) < cond(A)$. (c) Construct a 2×2 example where $||T|| > 1$ but $cond(I - T) \ll cond(A)$.*

Exercise 8.17. *(a) Find 2×2 examples where A SPD and M SPD does not imply $M^{-1}A$ is SPD. (b) Show however $M^{-1/2}AM^{-1/2}$ is SPD. (c) Show that if A or B is invertible then AB is similar to BA. Using this show that $M^{-1/2}AM^{-1/2}$ and has the same eigenvalues as $M^{-1}A$.*

Exercise 8.18. *Write down CG for $\left(M^{-1/2}AM^{-1/2}\right)y = M^{-1/2}b$. Reverse the change of variable everywhere $y \Leftarrow M^{1/2}x$ and eliminate all the $M^{\pm -1/2}$ to give the PCG algorithm as stated.*

Exercise 8.19. *In Exercise 8.1 (page 182), you wrote a program to apply CG to the 2D MPP. Modify that code to use PCG, Algorithm 8.8. Test* *it carefully, including one test using the identity matrix as preconditioner, making sure it results in exactly the same results as CG.*

Choose as preconditioner (a) the diagonal part of the matrix A (4 times the identity matrix); (b) the tridiagonal part of A (the tridiagonal matrix with 4 on the diagonal and -1 on the super- and sub-diagonals); and, (c) a third preconditioner of your choosing. Compare the numbers of iterations required and the total computer time required for each case.

8.6   CGN for Non-SPD Systems

> You must take your opponent into a deep dark forest where 2+2=5, and the path leading out is only wide enough for one.
> — Mikhail Tal

If A is SPD then the CG method is *provable the best possible one*. For a general linear system the whole beautiful structure of the CG method collapses. In the SPD case CG has the key properties that

- it is given by one 3 term (or two coupled 2 term) recursion,
- it has the finite termination property producing the exact solution in N steps for an $N \times N$ system,
- the nth step produces an approximate solution that is optimal over an n-dimensional affine subspace,
- it never breaks down,
- it takes at most $O(\sqrt{cond(A)})$ steps per significant digit.

There is a three-step, globally optimal, finite terminating CG method in the SPD case. In the non-symmetric case there is a fundamental result of Farber and Manteuffel and Voevodin [Faber and Manteuffel (1984)] and [Voevodin (1983)].

They proved a nonexistence theorem that no general extension of CG exists which retains these properties. The following is summary.

Theorem 8.10 (Faber, Manteuffel, and Voevodin). *Let A be an $N \times N$ real matrix. An s term conjugate gradient method exists for the matrix A if and only if either A is 3 by 3 or A is symmetric or A has a complete set of eigenvectors and the eigenvalues of A lie along a line in the complex plane.*

Thus, in a mathematically precise sense CG methods *cannot exist* for general nonsymmetric matrices. This means various extensions of CG to nonsymmetric systems seek to retain some of the above properties by giving up the others. Some generalized CG methods drop global optimality (and this means finite termination no longer holds) and some drop the restriction of a small recursion length (e.g., some have full recursions- the nth step has $k = n$). Since nothing can be a general best solution, there naturally have resulted many generalized CG methods which work well for some problems and poorly for others when A is nonsymmetric. (This fact by itself hints that none work well in all cases.) Among the popular ones today there are:

- **biCG** = biconjugate gradient method: biCG is based on an extension of the Orthogonalization of Moments algorithm to nonsymmetric matrices. It does not produce an orthogonal basis but rather two, a basis and a so-called shadow basis: $\{\phi_i : i = 1, \cdots, N\}$ and $\{\widetilde{\phi}_i : i = 1, \cdots, N\}$. The pair have the bi-orthogonality property that

$$\widetilde{\phi}_i^t A \phi_j = 0 \text{ for } i \neq j.$$

- **CGS** = conjugate gradient squared (which does not require A^t): CGS is an idea of Sonneveld that performed very well but resisted rigorous understanding for many years. Motivated by biCG, Sonneveld tried (loosely speaking) replacing the use of A^t by A in the algorithm wherever it occurred. This is of course very easy to test once biCG is implemented. The result was a method that converged in practice twice as fast as biCG.
- **GMRes** = generalized minimum residual method: GMRes was based on two modifications to CG. First the residual minimized at each step is $||b - Ax^{n+1}||_2^2$. This produces a method with no breakdowns at this step. Next orthogonalization of moments is replaced by the full Gram–Schmidt algorithm. The result is a memory expensive method which is optimal and does not break down.
- **CGNE and CGNR** = different realizations of CG for the normal equations

$$A^t A x = A^t b.$$

Of course an explicit or implicit change to the normal equations squares the condition number of the system being solved and requires A^t.

None in general work better than CGNE so we briefly describe CGNE. Again, we stress that for nonsymmetric systems, the "best" generalized CG

method will vary from one system to another. We shall restrict ourselves
to the case where A is square ($N \times N$). The following is known about the
normal equations.

Theorem 8.11 (The normal equations). *Let A be $N \times N$ and invertible. Then $A^t A$ is SPD. If A is SPD then*

$$cond_2(A^t A) = [cond_2(A)]^2 .$$

Proof. Symmetry: $(A^t A)^t = A^t A^{tt} = A^t A$. Positivity: $x^t(A^t A)x = (Ax)^t Ax = |Ax|^2 > 0$ for x nonzero since A is invertible. If A is SPD,
then $A^t A = A^2$ and

$$cond_2(A^t A) = cond_2(A^2) = \frac{\lambda_{\max}(A^2)}{\lambda_{\min}(A^2)}$$

$$= \frac{\lambda_{\max}(A)^2}{\lambda_{\min}(A)^2} = \left(\frac{\lambda_{\max}(A)}{\lambda_{\min}(A)}\right)^2$$

$$= [cond_2(A)]^2 .$$

$\square$

Thus, any method using the normal equation will pay a large price in
increasing condition numbers and numbers of iterations required. Beyond
that, if A is sparse, forming $A^t A$ directly shows that $A^t A$ will have roughly
double the number of nonzero entries per row as A. Thus, any algorithm
working with the normal equations avoids forming them explicitly. Residuals are calculated by multiplying by A and then multiplying that by A^t.

Algorithm 8.9 (CGNE = CG for the normal equations).
*Given preconditioner M, matrix A, initial vector x^0, right side vector b,
and maximum number of iterations* `itmax`

$r^0 = b - Ax^0$
$z^0 = A^t r^0$
$d^0 = z^0$
`for n=0:itmax`
    $\alpha_n = \langle \rho^n, \rho^n \rangle / \langle d^n, A^t(Ad^n) \rangle = \langle \rho^n, \rho^n \rangle / \langle Ad^n, Ad^n \rangle$
    $x^{n+1} = x^n + \alpha_n d^n,$
    $r^{n+1} = b - Ax^{n+1}$
    `if` *converged, exit,* `end`
    $z^{n+1} = A^t r^{n+1}$
    $\beta_{n+1} = \langle z^{n+1}, z^{n+1} \rangle / \langle z^n, z^n \rangle$

$$d^{n+1} = z^{n+1} + \beta_{n+1} d^n$$

end

Applying the convergence theory of CG, we have that CGN takes roughly the following number of steps per significant digit:

$$\frac{1}{2} \sqrt{cond_2(A^t A)} \simeq \frac{1}{2} \sqrt{[cond_2(A)]^2} \simeq \frac{1}{2} cond_2(A).$$

Since this is much larger than the SPD case of $\frac{1}{2}\sqrt{cond_2(A)}$ steps, preconditioning becomes much more important in the non SPD case than the SPD case. Naturally, much less is known in the non SPD case about construction of good preconditioners.

For the other variants, let us recall the good properties of CG as a way to discuss in general terms some of them. CG for SPD matrices A has the key properties that

- **For A SPD CG is given by one 3 term recursion or, equivalently, two coupled 2 term recursions:** This is an important property for efficiency. For A not SPD, GMRes drops this property and computes the descent directions by Gram–Schmidt orthogonalization. Thus for GMRes it is critical to start with a very good preconditioner and so limit the number of steps required. CGS retains a 3 term recursion for the search directions as does biCG and CGNE.
- **For A SPD it has the finite termination property producing the exact solution in at most N steps for an $N \times N$ system:** For A not SPD, biCG and full GMRes retain the finite termination property while CGS does not.
- **For A SPD the nth step produces an approximate solution that is optimal over a n-dimensional affine subspace:** For A not SPD, biCG and full GMRes retain this property while CGS does not.
- **For A SPD it never breaks down:** For A not SPD, breakdowns can occur. One method of dealing with them is to test for zero denominators and when one appears the algorithm is simply restarted taking the last approximation as the initial guess. biCG and CGS can have breakdowns. Full GMRes is reliable. Breakdowns can occur when, the full Gram–Schmidt orthogonalization procedure is truncated to a fixed number of steps.
- **For A SPD it takes at most $O(\sqrt{cond(A)})$ steps per significant digit:** For A not SPD, the question of the number of steps required

is very complex. On the one hard, one can phrase the question (indirectly) that if method X is applied and A happens to be SPD then, does method X reduce to CG? Among the methods mentioned, only biCG has this property. General (worst case) convergence results for these methods give no improvement over CGNE: they predict $O(cond_2(A))$ steps per significant digit. Thus the question is usually studied by computational tests which have shown that there are significant examples of nonsymmetric systems for which each of the methods mentioned is the best and requires significantly fewer than the predicted worst case number of steps.

Among the generalized CG methods for nonsymmetric systems, GMRes is the one currently most commonly used. It also seems likely that CGS is a method that is greatly under appreciated and under used.

Exercise 8.20. *The goal of this exercise is for you to design and analyze (reconstruct as much of the CG theory as you can) your own Krylov subspace iterative method that will possibly be better than CG. So consider solving $Ax = b$ where A is nonsingular. Given x^n, d^n the new iterate is computed by*

$$x^{n+1} = x^n + \alpha_n d^n$$
$$\alpha_n = \arg\min \|b - Ax^{n+1}\|_2^2.$$

(a) *Find a formula for α_n. Can this formula ever break down? Is there a zero divisor ever? Does the formula imply that x^{n+1} is a projection [best approximation] with respect to some inner product and norm? Prove it.*

(b) *Next consider your answer to part (a) carefully. Suppose the search directions are orthogonal with respect to this inner product. Prove a global optimality condition for your new method.*

(c) *What is the appropriate Krylov subspace to consider for the new method? Reconsider the Orthogonalization of Moments algorithm. Adapt it to give a algorithm and its proof for generating such an orthogonal basis.*

(d) *For this part you may choose: Either test the method and compare it with CG for various h's for the MPP or complete the error estimate for the method adapting the one for CG.*

Exercise 8.21. *Consider the non symmetric, 2×2 block system*

$$\begin{bmatrix} A_1 & C \\ -C^t & A_2 \end{bmatrix} \begin{bmatrix} x \\ y \end{bmatrix} = \begin{bmatrix} b_1 \\ b_2 \end{bmatrix}.$$

Suppose A_1 and A_2 are SPD and all blocks are $N \times N$. Check a 2×2 example that the eigenvalues of this matrix are not real. Consider preconditioning by the 2×2 block SPD system as follows:

$$\begin{bmatrix} A_1 & 0 \\ 0 & A_2 \end{bmatrix}^{-1} \begin{bmatrix} A_1 & C \\ -C^t & A_2 \end{bmatrix} \begin{bmatrix} x \\ y \end{bmatrix} = \begin{bmatrix} A_1 & 0 \\ 0 & A_2 \end{bmatrix}^{-1} \begin{bmatrix} b_1 \\ b_2 \end{bmatrix}.$$

Show that the eigenvalues of the preconditioned matrix

$$\begin{bmatrix} A_1 & 0 \\ 0 & A_2 \end{bmatrix}^{-1} \begin{bmatrix} A_1 & C \\ -C^t & A_2 \end{bmatrix}$$

lie on a line in the complex plane.

Chapter 9

Eigenvalue Problems

"Why is eigenvalue like liverwurst?"
— C. A. Cullen

9.1 Introduction and Review of Eigenvalues

By relieving the brain of all unnecessary work, a good notation sets it free to concentrate on more advanced problems, and, in effect, increases the mental power of the race.
— Whitehead, Alfred North (1861–1947), In P. Davis and R. Hersh, The Mathematical Experience, Boston: Birkhäuser, 1981.

One of the three fundamental problems of numerical linear algebra is to find information about the eigenvalues of an $N \times N$ matrix A. There are various cases depending on the structure of A (large and sparse vs. small and dense, symmetric vs. non-symmetric) and the information sought (the largest or dominant eigenvalue, the smallest eigenvalue vs. all the eigenvalues).

Definition 9.1 (eigenvalue-eigenvector). Let A be an $N \times N$ matrix. λ is an **eigenvalue** of A if there is a nonzero vector $\vec{\phi} \neq 0$ with

$$A\vec{\phi} = \lambda \vec{\phi}.$$

$\vec{\phi}$ is an **eigenvector** associated with the eigenvalue λ.

Eigenvalues are important quantities and often the figure of interest in physical problems. A few examples:

Vibration problems: Let $x(t) : [0, \infty) \to R^N$ satisfy

$$x''(t) + Ax(t) = 0.$$

For such problems vibratory or oscillatory motions at a fundamental frequency ω are critical to the observed dynamics of $x(t)$. Often the problem is to design a system (the design results in the matrix A) to fix its fundamental frequencies. Such an oscillatory solution takes the form

$$x(t) = \cos(\omega t)\, \overrightarrow{\phi}.$$

Inserting this into the ODE $x''(t) + Ax(t) = 0$ gives

$$-\omega^2 \cos(\omega t)\, \overrightarrow{\phi} + A\cos(\omega t)\, \overrightarrow{\phi} = 0$$

$$\Longleftrightarrow$$

$$A\overrightarrow{\phi} = \omega^2 \overrightarrow{\phi}.$$

Thus, ω is a fundamental frequency (and the resulting motion is nonzero and a persistent vibration) if and only if ω^2 is an eigenvalue of A. Finding fundamental frequencies means finding eigenvalues.

Buckling of a beam: The classic model for buckling of a thin beam is a yard stick standing and loaded on its top. Under light loads (and carefully balanced) it will stand straight. At a critical load it will buckle. The problem is to find the critical load. The linear elastic model for the displacement is the ODE

$$y''''(x) + \lambda y''(x) = 0, 0 < x < b,$$

$$y(0) = 0 = y''(0)$$

$$y(b) = 0 = y''(b).$$

The critical load can be inferred from the smallest value of λ for which the above has a nonzero solution.[1] If the derivatives in the ODE are replaced by difference quotients on a mesh, this leads to an eigenvalue problem for the resulting matrix. Finding the critical load under which buckling happens means finding an eigenvalue.

Stability of equilibria: If $x(t)$ is a function : $[0, \infty) \to R^N$ satisfies

$$x'(t) = F(x(t)),$$

$$F : R^N \to R^N$$

---

[1]This simple problem can be solved by hand using general solution of the ODE. If the problem becomes 1/2 step closer to a real problem from science or engineering, such as buckling of a 2D shell, it cannot be solved exactly. Then the only recourse is to discretize, replace it by an EVP for a matrix and solve that.

then an equilibrium solution is a vector x_0 satisfying $F(x_0) = 0$. If x(t) is another solution near the equilibrium solution, we can expand in a Taylor series near the equilibrium. The deviation from equilibrium $x(t) - x_0$ satisfies

$$(x(t) - x_0)' = F(x(t)) - F(x_0)$$
$$= F(x_0) + F'(x_0)(x(t) - x_0) + O(x(t) - x_0)^2$$
$$= F'(x_0)(x(t) - x_0) + (\text{small terms})^2.$$

Thus whether $x(t)$ approaches x_0 or not depends on the real parts of the eigenvalues of the $N \times N$ derivative matrix evaluated at the equilibrium. The equilibrium x_0 is locally stable provided the eigenvalues λ of $F'(x_0)$ satisfy $\text{Re}(\lambda) < 0$. Determining stability of rest states means finding eigenvalues.

Finding eigenvalues. Calculating λ, ϕ by hand (for small matrices) is a two step process which is simple in theory but seldom practicable.

Finding $\lambda, \vec{\phi}$ for A an $N \times N$ real matrix by hand:

- **Step 1:** Calculate exactly the characteristic polynomial of A. $p(\lambda) := \det(A - \lambda I)$ is a polynomial of degree N with real coefficients.
- **Step 2:** Find the N (counting multiplicities) real or complex roots of $p(\lambda) = 0$. These are the eigenvalues

$$\lambda_1, \lambda_2, \lambda_3, \cdots, \lambda_N.$$

- **Step 3:** For each eigenvalue λ_i, using Gaussian elimination find a non-zero solution of

$$[A - \lambda_i I] \vec{\phi}_i = 0, i = 1, 2, \cdots, N.$$

Example 9.1. Find the eigenvalues and eigenvectors of the 2×2 matrix

$$A = \begin{bmatrix} 1 & 1 \\ 4 & 1 \end{bmatrix}.$$

We calculate the degree 2 polynomial

$$p_2(\lambda) = \det(A - \lambda I) = \det \begin{bmatrix} 1 - \lambda & 1 \\ 4 & 1 - \lambda \end{bmatrix} = (1 - \lambda)^2 - 4.$$

Solving $p_2(\lambda) = 0$ gives

$$p_2(\lambda) = 0 \Leftrightarrow (1 - \lambda)^2 - 4 = 0 \Leftrightarrow \lambda_1 = 3, \lambda_2 = -1.$$

The eigenvector $\vec{\phi}_1$ of $\lambda_1 = 3$ is found by solving

$$(A - \lambda I) \begin{bmatrix} x \\ y \end{bmatrix} = \begin{bmatrix} -2 & 1 \\ 4 & -2 \end{bmatrix} \begin{bmatrix} x \\ y \end{bmatrix} = \begin{bmatrix} 0 \\ 0 \end{bmatrix}.$$

Solving gives (for any $t \in \mathbb{R}$)

$$y = t, \qquad -2x + y = 0, \text{ or } x = \frac{1}{2}t.$$

Thus, $(x, y)^t = (\frac{1}{2}t, t)^t$ for any $t \neq 0$ is an eigenvector. For example, $t = 2$ gives

$$\lambda_1 = +3, \ \overrightarrow{\phi}_1 = \begin{bmatrix} 1 \\ 2 \end{bmatrix}.$$

Similarly, we solve for $\overrightarrow{\phi}_2$

$$(A - \lambda I) \begin{bmatrix} x \\ y \end{bmatrix} = \begin{bmatrix} 2 & 1 \\ 4 & 2 \end{bmatrix} \begin{bmatrix} x \\ y \end{bmatrix} = \begin{bmatrix} 0 \\ 0 \end{bmatrix}$$

or $(x, y)^t = (-\frac{1}{2}t, t)^t$. Picking $t = 2$ gives

$$\lambda_2 = -1, \ \overrightarrow{\phi}_2 = \begin{bmatrix} -1 \\ 2 \end{bmatrix}.$$

Example 9.2 (An example of Wilkinson). The matrices

$$A = \begin{bmatrix} 2 & 1 \\ 0 & 2 \end{bmatrix},$$

$$A(\varepsilon) = \begin{bmatrix} 2 & 1 \\ -\varepsilon & 2 \end{bmatrix}$$

are close to each other for ε small. However their eigenspaces differ qualitatively. A has a double eigenvalue $\lambda = 2$ which only has 1 eigenvector. (The matrix A and eigenvalue $\lambda = 2$ are called defective.) The other matrix $A(\varepsilon)$ has distinct eigenvalues (and thus a complete set of eigenvectors)

$$\lambda_1(\varepsilon) = 2 + \sqrt{\varepsilon}$$
$$\lambda_2(\varepsilon) = 2 - \sqrt{\varepsilon}.$$

What is the sensitivity of the eigenvalues of A to perturbations? We calculate

$$\text{Sensitivity of } \lambda_i(\varepsilon) := \frac{d}{d\varepsilon} \lambda_i(\varepsilon)|_{\varepsilon=0}$$

$$= \pm \frac{1}{2} \varepsilon^{-1/2}|_{\varepsilon=0} = \infty.$$

Thus small changes of the coefficients of a defective matrix can produce large relative changes of its eigenvalues.

Exercise 9.1. *Analyze the sensitivity as $\varepsilon \to 0$ of the eigenvalues of the two matrices*

$$A = \begin{bmatrix} 2 & 0 \\ 0 & 2 \end{bmatrix}, \quad A(\varepsilon) = \begin{bmatrix} 2 & \varepsilon \\ 0 & 2 \end{bmatrix}.$$

Exercise 9.2. *Find the smallest in norm perturbation $(A \to A(\varepsilon))$ of the 2×2 diagonal matrix $A = diag(\lambda_1, \lambda_2)$ that merges its eigenvalues to a double eigenvalue of $A(\varepsilon)$ having value $(\lambda_1 + \lambda_2)/2$.*

9.1.1 *Properties of eigenvalues*

Some important properties of eigenvalues and eigenvectors are given below.

Proposition 9.1. *If A is diagonal, upper triangular or lower triangular, then the eigenvalues are on the diagonal of A.*

Proof. Let A be upper triangular. Then, using $*$ to denote a generic non-zero entry,

$$\det [A - \lambda I] = \det \begin{bmatrix} a_{11} - \lambda & * & * & * \\ 0 & a_{22} - \lambda & * & * \\ 0 & 0 & \ddots & * \\ 0 & 0 & 0 & a_{nn} - \lambda \end{bmatrix}$$

$$= (a_{11} - \lambda)(a_{22} - \lambda) \cdot \ldots \cdot (a_{nn} - \lambda) = p_n(\lambda).$$

The roots of p_n are obvious! $\qquad\square$

Proposition 9.2. *Suppose A is $N \times N$ and $\lambda_1, \ldots, \lambda_k$ are distinct eigenvalues of A. Then,*

$$\det(A - \lambda I) = (\lambda_1 - \lambda)^{p_1}(\lambda_2 - \lambda)^{p_2} \cdot \ldots \cdot (\lambda_k - \lambda)^{p_k}$$

where $p_1 + p_2 + \ldots + p_k = n$.

Each λ_j has at least one eigenvector ϕ_j and possibly as many as p_k linearly independent eigenvectors.

If each λ_j has p_j linearly independent eigenvectors then all the eigenvectors together form a basis for $\mathbb{R}^N$.

Proposition 9.3. *If A is symmetric (and real) $(A = A^t)$, then:*
(i) all the eigenvalues and eigenvectors are real,

(ii) there exists N orthonormal[2] eigenvectors $\overrightarrow{\phi}_1, \ldots, \overrightarrow{\phi}_N$ of A:

$$\langle \overrightarrow{\phi}_i, \overrightarrow{\phi}_j \rangle = \begin{cases} 1, \text{ if } i = j, \\ 0, \text{ if } i \neq j. \end{cases}$$

(iii) if C is the $N \times N$ matrix with eigenvector $\overrightarrow{\phi}_j$ in the jth column then

$$C^{-1} = C^t \quad \text{and} \quad C^{-1}AC = \begin{bmatrix} \lambda_1 & 0 & \ldots & 0 \\ 0 & \lambda_2 & \ldots & 0 \\ \vdots & \vdots & \ddots & \vdots \\ 0 & 0 & \ldots & \lambda_N \end{bmatrix}.$$

Proposition 9.4. *If an eigenvector $\overrightarrow{\phi}$ is known, the corresponding eigenvalue is given by the Rayleigh quotient*

$$\lambda = \frac{\overrightarrow{\phi}^* A \overrightarrow{\phi}}{\overrightarrow{\phi}^* \overrightarrow{\phi}}, \quad \text{where } \overrightarrow{\phi}^* = \text{ conjugate transpose } = \overline{\left(\overrightarrow{\phi}\right)}^{tr}.$$

Proof. If $A\overrightarrow{\phi} = \lambda \overrightarrow{\phi}$, we have

$$\overrightarrow{\phi}^* A \overrightarrow{\phi} = \lambda \overrightarrow{\phi}^* \overrightarrow{\phi}$$

from which the formula for λ follows.   □

Proposition 9.5. *If $\| \cdot \|$ is any matrix norm (induced by the vector norm $\| \cdot \|$),*

$$|\lambda| \leq \|A\|.$$

Proof. Since

$$\lambda \overrightarrow{\phi} = A \overrightarrow{\phi}$$

we have

$$|\lambda| \| \overrightarrow{\phi} \| = \| \lambda \overrightarrow{\phi} \| = \| A \overrightarrow{\phi} \| \leq \|A\| \| \overrightarrow{\phi} \|.$$

    □

Remark 9.1. The eigenvalues of A are complicated, nonlinear functions of the entries in A. Thus, the eigenvalues of A+B have no correlation with those of A and B. In general, $\lambda(A + B) \neq \lambda(A) + \lambda(B)$.

---

[2] "Orthonormal" means orthogonal (meaning mutually perpendicular so their dot products give zero) and normal (meaning their length is normalized to be one).

9.2 Gershgorin Circles

The question we consider in this section is:

What can we tell about the eigenvalues of A from the entries in the matrix A?

Eigenvalues are very important yet they are complicated, nonlinear functions of the entries of A. Thus, results that allow us to look at the entries of A and get information about where the eigenvalues live are useful results indeed. We have seen two already for $N \times N$ real matrices:

- If $A = A^t$ then $\lambda(A)$ is real (and, by a similar argument, if $A^t = -A$ then the eigenvalues are purely imaginary).
- $|\lambda| \leq \|A\|$ for any norm $\| \cdot \|$; in particular this means $|\lambda| \leq \min\{\|A\|_1, \|A\|_\infty\}$.

Definition 9.2. The spectrum of A, $\sigma(A)$, is
$$\sigma(A) := \{\lambda | \lambda \text{ is an eigenvalue of } A\}.$$
The *numerical range* $R(A) := \{x^* A x | \text{ for all } x \in \mathbb{C}^N, \|x\|_2 = 1\}$.

This question is often called *"spectral localization"*. The two classic spectral localization results are $\sigma(A) \subset R(A)$ and the Gershgorin circle theorem.

Theorem 9.1 (Properties of Numerical Range). *For* A *an* $N \times N$ *matrix,*

- $\sigma(A) \subset R(A)$
- $R(A)$ *is compact and convex (and hence simply connected).*
- *If* A *is normal matrix (i.e.,* A *commutes with* A^t*) then* $R(A)$ *is the convex hull of* $\sigma(A)$.

Proof. The second claim is the celebrated Toeplitz-Hausdorff theorem. We only prove the first claim. Picking $x =$ the eigenvector of λ of unit length gives: $x^* A x =$ the eigenvalue λ. $\qquad\square$

Theorem 9.2 (The Gershgorin Circle Theorem). *Let* $A = (a_{ij})$, i, $j = 1, \ldots, N$. *Define the row and column sums which exclude the diagonal entry.*

$$r_k = \sum_{j=1, j \neq k}^{N} |a_{kj}|, \qquad c_k = \sum_{j=1, j \neq k}^{N} |a_{jk}|.$$

Define the closed disks in $\mathbb{C}$:

$$R_k = \{z \in \mathbb{C};\ |z - a_{kk}| \le r_k\}, \qquad C_k = \{z \in \mathbb{C};\ |z - a_{kk}| \le c_k\}.$$

Then, if λ *is an eigenvalue of* A

(1) $\lambda \in R_k$ *for some* k.
(2) $\lambda \in C_k$ *for some* k.
(3) *If* Ω *is a union of precisely* k *disks that is disjoint from all other disks then* Ω *must contain* k *eigenvalues of* A.

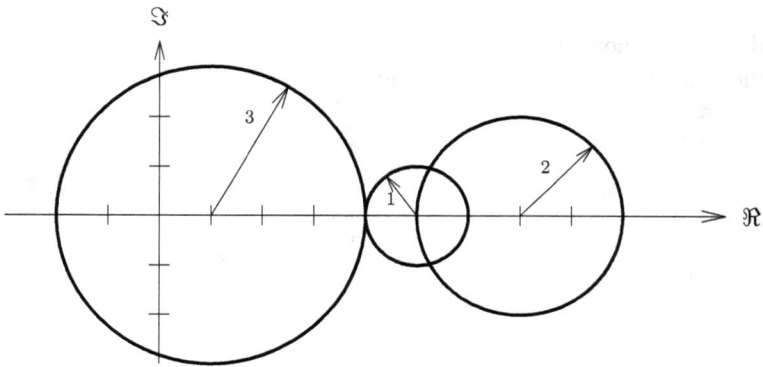

Fig. 9.1   Three Gershgorin disks in Example 9.3.

Example 9.3.

$$A_{3\times 3} = \begin{bmatrix} 1 & 2 & -1 \\ 2 & 7 & 0 \\ -1 & 0 & 5 \end{bmatrix}.$$

We calculate

$$r_1 = 2 + 1 = 3, \qquad r_2 = 2 + 0 = 2, \qquad r_3 = 1 + 0 = 1.$$
$$c_1 = 2 + 1 = 3, \qquad c_2 = 2 + 0 = 2, \qquad c_k = 1 + 0 = 1.$$

The eigenvalues must belong to the three disks in Figure 9.1. Since $A = A^t$, they must also be real. Thus

$$-2 \le \lambda \le 8.$$

Exercise 9.3. *If* B *is a submatrix of* A *(constructed by deleting the same number of rows and columns), show that* $R(B) \subset R(A)$.

9.3 Perturbation Theory of Eigenvalues

Whatever method is used to calculate approximate eigenvalues, in finite precision arithmetic what is actually calculated are the eigenvalues of a nearby matrix. Thus, the first question is "How are the eigenvalues changed under small perturbations?" It is known that the eigenvalues of a matrix are continuous functions of the entries of a matrix. However, the modulus of continuity can be large; small changes in some matrices can produce large changes in the eigenvalues. One class of matrices that are well conditioned with respect to its eigenvalues is real symmetric matrices.

Example 9.4 (An example of Forsythe). Let A, E be the $N \times N$ matrices: for $a > 0$ and $\varepsilon > 0$ small:

$$
A = \begin{bmatrix} a & 0 & 0 & \cdots & 0 \\ 1 & a & 0 & \cdots & 0 \\ 0 & 1 & a & & 0 \\ \vdots & & \ddots & \ddots & \vdots \\ 0 & \cdots & 0 & 1 & a \end{bmatrix},
$$

$$
E = \begin{bmatrix} 0 & 0 & \cdots & 0 & \varepsilon \\ 0 & 0 & \cdots & 0 & 0 \\ 0 & 0 & \cdots & 0 & 0 \\ & & & & \\ 0 & 0 & \cdots & 0 & 0 \end{bmatrix}.
$$

Then the characteristic equations of A and A+E are, respectively,

$$
(a - \lambda)^N = 0 \quad \text{and} \quad (a - \lambda)^N + \varepsilon(-1)^{N+1} = 0.
$$

Thus, the eigenvalues of A are $\lambda_k = a, a, a, \cdots$ while those of A+E are

$$
\mu_k = a + \omega^k \varepsilon^{1/N}, k = 0, 1, \cdots, N - 1
$$

where ω is a primitive Nth root of unity. Thus $A \to A + E$ changes one, multiple, real eigenvalue into N distinct complex eigenvalues about a with radius $\varepsilon^{1/N}$. Now suppose that

- $\varepsilon = 10^{-10}, N = 10$ then the error in the eigenvalues is 0.1. *It has been magnified by 10^9!*
- $\varepsilon = 10^{-100}, N = 100$ then the error in the eigenvalues is $0.1 = 10^{99} \times$ Error in A!

9.3.1 *Perturbation bounds*

For simplicity, suppose A is diagonalizable

$$H^{-1}AH = \Lambda = diag(\lambda_1, \lambda_2, \cdots, \lambda_N)$$

and let λ, μ denote the eigenvalues of A and $A + E$ respectively:

$$Ax = \lambda x, \quad (A + E)y = \mu y.$$

Theorem 9.3. *Let A be diagonalizable by the matrix H and let λ_i denote the eigenvalues of A. Then, for each eigenvalue μ of $A + E$,*

$$\min_i |\mu - \lambda_i| \leq ||H||_2||H^{-1}||_2||E||_2 = cond_2(H)||E||_2$$

Proof. The eigenvector y of $A + E$ satisfies the equation

$$(\mu I - A)y = Ey.$$

If μ is an eigenvalue of A the the result holds since the LHS is zero. Otherwise, we have

$$H^{-1}(\mu I - A)HH^{-1}y = H^{-1}EHH^{-1}y, \quad \text{or}$$
$$(\mu I - \Lambda)w = \left(H^{-1}EH\right)w, \quad \text{where} \quad w = H^{-1}y.$$

In this case $(\mu I - \Lambda)$ is invertible. Thus

$$w = (\mu I - \Lambda)^{-1}\left(H^{-1}EH\right)w.$$

So that

$$||w||_2 \leq ||(\mu I - \Lambda)^{-1}||_2||H^{-1}||_2||E||_2||H||_2||w||_2.$$

The result follows since

$$||(\mu I - \Lambda)^{-1}||_2 = \max_i \frac{1}{|\mu - \lambda_i|} = \left(\min_i |\mu - \lambda_i|\right)^{-1}.$$

$$\square$$

Definition 9.3. Let λ, μ denote the eigenvalues of A and $A + E$ respectively. The eigenvalues of the real, symmetric matrix A are called "well conditioned" when

$$\min_i |\mu - \lambda_i| \leq ||E||_2.$$

Proof. In this case note that H is orthogonal and thus $||H||_2 = ||H^{-1}||_2 = 1$.

$$\square$$

Other results are known such as:

Theorem 9.4. *Let A be real and symmetric. There is an ordering of eigenvalues of A and $A + E$ under which*

$$\max_i |\mu_i - \lambda_i| \leq ||E||_2,$$

$$\sum_{i=1,N} |\mu_i - \lambda_i|^2 \leq ||E||_{Frobenius}^2. \tag{9.1}$$

For more information see the book of Wilkinson.

J. WILKINSON, *The Algebra Eigenvalue Problem*, Oxford Univ. Press, 1965.

9.4 The Power Method

The power method is used to find the dominant (meaning the largest in complex modulus) eigenvalue of a matrix A. It is specially appropriate when A is large and sparse so multiplying by A is cheap in both storage and in floating point operations. If a complex eigenvalue is sought, then the initial guess in the power method must also be complex. In this case the inner product of complex vectors is the conjugate transpose:

$$\langle x, y \rangle := x^* y := \bar{x}^T y = \sum_{i=1}^{N} \bar{x}_i y_i.$$

Algorithm 9.1 (Power Method for Dominant Eigenvalue).
Given a matrix A, an initial vector $x^0 \neq 0$, and a maximum number of iterations itmax,

    for n=0:itmax
        $\tilde{x}^{n+1} = A x^n$
        $x^{n+1} = \tilde{x}^{n+1} / ||\tilde{x}^{n+1}||$
        % *estimate the eigenvalue by*
        $(*)$ $\lambda = (x^{n+1})^* A x^{n+1}$
        if *converged, stop,* end
    end

Remark 9.2. The step $(*)$ in which the eigenvalue is recovered can be rewritten as

$$\lambda_{n+1} = (\tilde{x}^{n+2})^* x^{n+1}$$

since $\tilde{x}^{n+2} = A x^{n+1}$. Thus it can be computed without additional cost.

9.4.1   *Convergence of the power method*

In this section we examine convergence of the power method for the case that the dominant eigenvalue is simple. First we note that the initial guess must have some component in the direction of the eigenvector of the dominant eigenvalue. We shall show that the power method converges rapidly when the dominant eigenvalue is well separated from the rest:

$$|\lambda_1| \gg |\lambda_j|, \qquad j = 2, \ldots, N.$$

In order to illuminate the basic idea, we shall analyze its convergence under the additional assumption that A has N linearly independent eigenvectors and that the dominant eigenvector is simple and real.

With $x^0 \in \mathbb{R}^N$ and eigenvectors $\vec{\phi}_1, \vec{\phi}_2$ of A ($A\vec{\phi}_j = \lambda_j \vec{\phi}_j$) we can expand the initial guess in terms of the eigenvectors of A as follows:

$$x^0 = c_1 \vec{\phi}_1 + c_2 \vec{\phi}_2 + \ldots + c_N \vec{\phi}_N.$$

If the initial guess has some component in the first eigenspace then

$$c_1 \neq 0.$$

Then we calculate the normalized[3] iterates $\widetilde{x}^1 = A\widetilde{x}^0, \widetilde{x}^2 = A\widetilde{x}^1 = A^2\widetilde{x}^0$, etc.:

$$\widetilde{x}^1 = A\widetilde{x}^0 = c_1 A\vec{\phi}_1 + \ldots + c_n A\vec{\phi}_n$$
$$= c_1\lambda_1 \vec{\phi}_1 + c_2\lambda_2 \vec{\phi}_2 + \ldots + c_N\lambda_N \vec{\phi}_N,$$

$$\widetilde{x}^2 = A\widetilde{x}^1 = c_1\lambda_1 A\vec{\phi}_1 + \ldots + c_n\lambda_N A\vec{\phi}_n$$
$$= c_1\lambda_1^2 \vec{\phi}_1 + c_2\lambda_2^2 \vec{\phi}_2 + \ldots + c_N\lambda_N^2 \vec{\phi}_N,$$

$$\vdots$$

$$\widetilde{x}^k = A\widetilde{x}^{k-1} = A^{k-1}\widetilde{x}^0$$
$$= c_1\lambda_1^k \vec{\phi}_1 + c_2\lambda_2^k \vec{\phi}_2 + \ldots + c_N\lambda_N^k \vec{\phi}_N.$$

Since $|\lambda_1| > |\lambda_j|$ the largest contribution to $||\widetilde{x}^k||$ is the first term. Thus, normalize $\widetilde{x}^k$ by the size of the first term so that

$$\frac{1}{\lambda_1^k}\widetilde{x}^k = c_1 \vec{\phi}_1 + c_2 \left(\frac{\lambda_2}{\lambda_1}\right)^k \vec{\phi}_2 + \ldots + c_N \left(\frac{\lambda_N}{\lambda_1}\right)^k \vec{\phi}_N.$$
$$\downarrow \qquad\qquad \downarrow \qquad\quad \downarrow$$
$$0 \qquad\qquad 0 \qquad\quad 0$$

---

[3]The $\widetilde{x}$ here are different from those in the algorithm because the normalization is different.

Each term except the first $\to 0$ since $\left|\frac{\lambda_2}{\lambda_1}\right| < 1$. Thus,

$$\frac{1}{\lambda_1^k}\tilde{x}^k = c_1\vec{\phi}_1 + (\text{ terms that } \to 0 \text{ as } k \to \infty),$$

or,

$$\tilde{x}^k \simeq c_1\lambda_1^k\vec{\phi}_1, \quad \vec{\phi}_1 = \text{ eigenvector of } \lambda_1,$$

so $A\tilde{x}^k \simeq A(c_1\lambda_1^k\vec{\phi}_1) = c_1\lambda_1^{k+1}\vec{\phi}_1$ or

$$A\tilde{x}^k \simeq \lambda_1\tilde{x}^k$$

and so we have found $\lambda_1, \vec{\phi}_1$ approximately.

Example 9.5. $A = \begin{bmatrix} 2 & 4 \\ 3 & 13 \end{bmatrix}$, $x^0 = \begin{bmatrix} 1 \\ 0 \end{bmatrix}$. Then

$$\tilde{x}^1 = Ax^0 = \begin{bmatrix} 2 & 4 \\ 3 & 13 \end{bmatrix}\begin{bmatrix} 1 \\ 0 \end{bmatrix} = \begin{bmatrix} 2 \\ 3 \end{bmatrix},$$

$$x^1 = \frac{\tilde{x}^1}{\|\tilde{x}^1\|} = \frac{[2,3]^t}{\sqrt{2^2+3^2}} = \begin{bmatrix} .5547 \\ .8321 \end{bmatrix}.$$

$$\tilde{x}^2 = Ax^1 = \begin{bmatrix} 4.438 \\ 12.48 \end{bmatrix}.$$

$$x^2 = \frac{\tilde{x}^2}{\|\tilde{x}^2\|} = \ldots, \quad \text{and so on.}$$

Exercise 9.4. *Write a computer program to implement the power method, Algorithm 9.1. Regard the algorithm as converged when $|Ax^{n+1} - \lambda x^{n+1}| < \epsilon$, with $\epsilon = 10^{-4}$. Test your program by computing $\tilde{x}^1$, x^1 and $\tilde{x}^2$ in Example 9.5 above. What are the converged eigenvalue and eigenvector in this example? How many steps did it take?*

9.4.2 *Symmetric matrices*

The Power Method converges twice as fast for symmetric matrices as for non-symmetric matrices because of some extra error cancellation that occurs due to the eigenvalues of symmetric matrices being orthogonal. To see this, suppose $A = A^t$ and calculate

$$\tilde{x}^{k+1} = A\tilde{x}^k (= \ldots = A^k x^0).$$

Then the kth approximation to λ is μ^k given by

$$\mu^k = \frac{(\tilde{x}^k)^t A\tilde{x}^k}{(\tilde{x}^k)^t\tilde{x}^k} = (\tilde{x}^k)^t\tilde{x}^{k+1}/(\tilde{x}^k)^t\tilde{x}^k.$$

If $x^0 = c_1 \vec{\phi}_1 + c_2 \vec{\phi}_2 + \ldots + c_N \vec{\phi}_N$ then, as in the previous case

$$\tilde{x}^k = c_1 \lambda_1^k \vec{\phi}_1 + \ldots + c_N \lambda_N^k \vec{\phi}_N, \quad \text{and thus}$$

$$\tilde{x}^{k+1} = c_1 \lambda_1^{k+1} \vec{\phi}_1 + \ldots + c_N \lambda_N^{k+1} \vec{\phi}_N.$$

In the symmetric case the eigenvectors are mutually orthogonal:

$$\vec{\phi}_i^t \vec{\phi}_j = 0, i \neq j.$$

Using orthogonality we calculate

$$(x^k)^t x^{k+1}$$

$$= \left(c_1 \lambda_1^k \vec{\phi}_1 + \ldots + c_N \lambda_N^k \vec{\phi}_N \right)^t \left(c_1 \lambda_1^{k+1} \vec{\phi}_1 + \ldots + c_N \lambda_N^{k+1} \vec{\phi}_N \right)$$

$$= \ldots = c_1^2 \lambda_1^{2k+1} + c_2^2 \lambda_2^{2k+1} + \ldots + c_N^2 \lambda_N^{2k+1}.$$

Similarly

$$(x^k)^t x^k = c_1^2 \lambda_1^{2k} + \ldots + c_N^2 \lambda_N^{2k}$$

and we find

$$\mu^k = \frac{c_1^2 \lambda_1^{2k+1} + \ldots + c_N^2 \lambda_N^{2k+1}}{c_1^2 \lambda_1^{2k} + \ldots + c_N^2 \lambda_N^{2k}}$$

$$= \frac{c_1^2 \lambda_1^{2k+1}}{c_1^2 \lambda_1^{2k}} + O\left(\left| \frac{\lambda_2}{\lambda_1} \right|^{2k} \right)$$

$$= \lambda_1 + O\left(\left| \frac{\lambda_2}{\lambda_1} \right|^{2k} \right),$$

which is twice as fast as the non-symmetric case!

Exercise 9.5. *Take $A_{2 \times 2}$ given below. Find the eigenvalues of A. Take $x^0 = (1, 2)^t$ and do 2 steps of the power method. If it is continued, to which eigenvalue will it converge? Why?*

$$A = \begin{bmatrix} 2 & -1 \\ -1 & 2 \end{bmatrix}.$$

9.5 Inverse Power, Shifts and Rayleigh Quotient Iteration

The idea behind variants of the power method is to replace A by a matrix whose largest eigenvalue is the one sought, find that by the power method for the modified matrix and then recover the sought eigenvalue of A.

9.5.1 *The inverse power method*

Although this may seem a paradox, all exact science is dominated by the idea of approximation.
— Russell, Bertrand (1872–1970), in W. H. Auden and L. Kronenberger (eds.) "The Viking Book of Aphorisms", New York: Viking Press, 1966.

The *inverse power method* computes the eigenvalue of A closest to the origin — the smallest eigenvalue of A. The inverse power method is equivalent to the power method applied to A^{-1} (since the smallest eigenvalue of A is the largest eigenvalue of A^{-1}).

Algorithm 9.2 (Inverse power method). *Given a matrix A, an initial vector $x^0 \neq 0$, and a maximum number of iterations* itmax,

for n=0:itmax
   (∗) *Solve* $A\widetilde{x}^{n+1} = x^n$
   $x^{n+1} = \widetilde{x}^{n+1}/\|\widetilde{x}^{n+1}\|$
   if *converged*, **break, end**
end
% *The converged eigenvalue is given by*
$\mu = (\widetilde{x}^{n+1})^* x^n$
$\lambda = 1/\mu$

For large sparse matrices step (∗) is done by using some other iterative method for solving a linear system with coefficient matrix A. Thus the total cost of the inverse power method is:

(number of steps of the inverse power method)*(number of iterations per step required to solve $A\widetilde{x}^{n+1} = x^n$)

This product can be large. Thus various ways to accelerate the inverse power method have been developed. Since the number of steps depends on the separation of the dominant eigenvalue from the other eigenvalues, most methods do this by using shifts to get further separation. If α is fixed, then the largest eigenvalue of $(A - \alpha I)^{-1}$ is related to the eigenvalue of A closest to α, λ_α by

$$\lambda_{\max}(A - \alpha I) = \frac{1}{\lambda_\alpha(A) - \alpha}.$$

The inverse power method with shift finds the eigenvalue closest to α.

Algorithm 9.3 (Inverse power method with shifts). *Given a matrix A, an initial vector $x^0 \neq 0$, a shift α, and a maximum number of iterations* `itmax`,

```
for n=0:itmax
    (*)   Solve (A − αI)x̃ⁿ⁺¹ = xⁿ
    xⁿ⁺¹ = x̃ⁿ⁺¹/‖x̃ⁿ⁺¹‖
    if converged, break, end
end
```
% *The converged eigenvalue is given by*
$\mu = (\widetilde{x}^{n+1})^* x^n$
$\lambda = \alpha + 1/\mu$

9.5.2 Rayleigh Quotient Iteration

> Fourier is a mathematical poem.
> — Thomson, [Lord Kelvin] William (1824–1907)

The Power Method and the Inverse Power Method are related to (and combine to form) Rayleigh Quotient Iteration. Rayleigh Quotient Iteration finds very quickly the eigenvalue closest to the initial shift for symmetric matrices. It is given by:

Algorithm 9.4 (Rayleigh Quotient Iteration). *Given a matrix A, an initial vector $x^0 \neq 0$, an initial eigenvalue λ^0, and a maximum number of iterations* `itmax`,

```
for n=0:itmax
    (*)   Solve (A − λⁿI)x̃ⁿ⁺¹ = xⁿ
    xⁿ⁺¹ = x̃ⁿ⁺¹/‖x̃ⁿ⁺¹‖
    λⁿ⁺¹ = (xⁿ⁺¹)ᵗAxⁿ⁺¹
    if converged, return, end
end
```

It can be shown that for symmetric matrices

$$\|x^{n+1} - \vec{\phi}_1\| \leq C\|x^n - \vec{\phi}_1\|^3,$$

i.e., the number of significant digits triples at each step in Rayleigh quotient iteration.

Remark 9.3. The matrix $A - \lambda^n I$ will become ill-conditioned as the iteration converges and λ^n approaches an eigenvalue of A. This ill-conditioning helps the iteration rather than hinders it because roundoff errors accumulate fastest in the direction of the eigenvector.

9.6   The QR Method

"But when earth had covered this generation also, Zeus the son of Cronos made yet another, the fourth, upon the fruitful earth, which was nobler and more righteous, a god-like race of hero-men who are called demi-gods, the race before our own, throughout the boundless earth. Grim war and dread battle destroyed a part of them, some in the land of Cadmus at seven-gated Thebe when they fought for the flocks of Oedipus, and some, when it had brought them in ships over the great sea gulf to Troy for rich-haired Helen's sake: there death's end enshrouded a part of them. But to the others father Zeus the son of Cronos gave a living and an abode apart from men, and made them dwell at the ends of earth. And they live untouched by sorrow in the islands of the blessed along the shore of deep swirling Ocean, happy heroes for whom the grain-giving earth bears honey-sweet fruit flourishing thrice a year, far from the deathless gods..."
— Hesiod, Works and Days

The QR algorithm is remarkable because if A is a small, possibly dense matrix the algorithm gives a reliable calculation of *all* the eigenvalues of A. The algorithm is based on the observation that the proof of existence of a QR factorization is constructive. First we recall the theorem of existence.

Theorem 9.5. *Let A be an $N \times N$ matrix. Then there exists*

- *a unitary matrix Q and*
- *an upper triangular matrix R*

such that

$$A = QR.$$

Moreover, R can be constructed so that the diagonal entries satisfy $R_{ii} \geq 0$. If A is invertible then there is a unique factorization with $R_{ii} \geq 0$.

Proof. Sketch of proof: Suppose A is invertible then the columns of A span $\mathbb{R}^N$. Let a_i, q_i denote the column vectors of A and Q respectively. With $R = r_{ij}$ upper triangular, writing out the equation $A = QR$ gives the following system:

$$a_1 = r_{11}q_1$$
$$a_2 = r_{12}q_1 + r_{22}q_2$$
$$\cdots$$
$$a_N = r_{1n}q_1 + r_{2n}q_2 + \cdots + r_{NN}q_N.$$

Thus, in this form, the QR factorization takes a spanning set $a_i, i = 1, \cdots, N$ and from that constructs an orthogonal set $q_i, i = 1, \cdots, N$ with

$$span\{a_1, a_2, \cdots, a_k\} = span\{q_1, q_2, \cdots, q_k\} \text{ for every } k.$$

Thus the entries in R are just the coefficients generated by the Gram–Schmidt process! This proves existence when A is invertible. $\square$

Remark 9.4. We remark that the actual calculation of the QR factorization is done stably by using Householder transformations rather than Gram–Schmidt.

The QR algorithm to calculate eigenvalues is built upon repeated construction of QR factorizations. Its cost is

$$\text{cost of the QR algorithm} \simeq 1 \text{ to } 4 \; LU \text{ decompositions} \simeq \tfrac{4}{3}N^3 \text{ FLOPs.}$$

Algorithm 9.5 (Simplified QR algorithm). *Given a square matrix A_1 and a maximum number of iterations* `itmax`,

```
for n=1:itmax
    Factor Aₙ = QₙRₙ
    Form Aₙ₊₁ = RₙQₙ
    if converged, return, end
end
```

This algorithm converges in an unusual sense:

- A_k is similar to A_1 for every k, and
- $(A_k)_{ij} \to 0$ for $i > j$,
- $diagonal(A_k) \to$ eigenvalues of A,
- $(A_k)_{ij}$ for $i < j$ does not necessarily converge to anything.

Various techniques are used to speed up convergence of QR such as using shifts.

Exercise 9.6. *Show that A_k is similar to A_1 for every k.*

Exercise 9.7. *Show that if A is real and $A = QR$ then so are Q and R.*

Exercise 9.8. *Show $A = QDS$ for invertible A with Q unitary and D diagonal and positive and S upper triangular with diagonal entries all 1.*

Exercise 9.9. *Show that if $A = QR$ and A is real then so are Q and R.*

Exercise 9.10. *Find the QR factorization of*

$$\begin{bmatrix} 2 & -1 & 0 \\ -1 & 2 & -1 \\ 0 & -1 & 2 \end{bmatrix}.$$

For more information see [Watkins (1982)].

Angling may be said to be so like mathematics that it can never be fully learned.
— Walton, Izaak, The Compleat Angler, 1653.

Appendix A

An Omitted Proof

Whatever regrets may be, we have done our best.
— Sir Ernest Shackleton, January 9, 1909, 88° 23′ South

The proof of Theorem 6.4 depends on the well-known Jordan canonical form of the matrix.

Theorem A.1 (Jordan canonical form). *Given a $N \times N$ matrix T, an invertible matrix C can be found so that*

$$T = CJC^{-1}$$

where J is a block diagonal matrix

$$J = \begin{bmatrix} J_1 & & & \\ & J_2 & & \\ & & \ddots & \\ & & & J_K \end{bmatrix} \tag{A.1}$$

and each of the $n_i \times n_i$ diagonal blocks J_i has the form

$$J_i = \begin{bmatrix} \lambda_i & 1 & & \\ & \lambda_i & 1 & \\ & & \ddots & \ddots \\ & & & \lambda_i & 1 \\ & & & & \lambda_i \end{bmatrix} \tag{A.2}$$

where λ_i is an eigenvalue of T. The λ_i need not be distinct eigenvalues.

The proof of this theorem is beyond the scope of this text, but can be found in any text including elementary linear algebra, such as the beautiful book of Herstein [Herstein (1964)].

Theorem 6.4 is restated here:

Theorem A.2. *Given any $N \times N$ matrix T and any $\varepsilon > 0$ there exists a matrix norm $\| \cdot \|$ with $\|T\| \le \rho(T) + \varepsilon$.*

Proof. Without loss of generality, assume $\varepsilon < 1$. Consider the matrix

$$
E_\varepsilon = \begin{bmatrix}
1 & & & & \\
& 1/\varepsilon & & & \\
& & 1/\varepsilon^2 & & \\
& & & \ddots & \\
& & & & 1/\varepsilon^{N-1}
\end{bmatrix}
$$

and the product $E_\varepsilon J E_\varepsilon^{-1}$. The first block of this product can be seen to be

$$
\begin{bmatrix}
\lambda_1 & \varepsilon & & & \\
& \lambda_1 & \varepsilon & & \\
& & \ddots & \ddots & \\
& & & \lambda_1 & \varepsilon \\
& & & & \lambda_1
\end{bmatrix}
$$

and each of the other blocks is similar. It is clear that $\|E_\varepsilon J E_\varepsilon^{-1}\|_\infty \le \rho(T) + \varepsilon$. Defining the norm $\|T\| = \|E_\varepsilon J E_\varepsilon^{-1}\|_\infty$ completes the proof. $\quad\square$

Remark A.1. If it happens that each of the eigenvalues with $|\lambda_i| = \rho(T)$ is *simple*, then each of the corresponding Jordan blocks is 1×1 and $\|T\| = \rho(T)$.

Appendix B

Tutorial on Basic MATLAB Programming

...Descartes, a famous philosopher, author of the celebrated dictum, *Cogito ergo sum* — whereby he was pleased to suppose he demonstrated the reality of human existence. The dictum might be improved, however, thus: *Cogito cogito ergo cogito sum* — "I think that I think, therefore I think that I am"; as close an approach to certainty as any philosopher has yet made.

— Ambrose Bierce, "The Devil's Dictionary"

B.1   Objective

The purpose of this appendix is to introduce the reader to the basics of the MATLAB programming (or scripting) language. By "basics" is meant the basic syntax of the language for arithmetical manipulations. The intent of this introduction is twofold:

(1) Make the reader sufficiently familiar with MATLAB that the pseudocode used in the text is transparent.
(2) Provide the reader with sufficient syntactical detail to expand pseudocode used in the text into fully functional programs.

In addition, pointers to some of the very powerful MATLAB functions that implement the algorithms discussed in this book are given.

The MATLAB language was chosen because, at the level of detail presented here, it is sufficiently similar to other languages such as C, C++, Fortran, and Java, that knowledge of one can easily be transferred to the others. Except for a short discussion of array syntax and efficiency, all of the programming constructs discussed in this appendix can be simply[1] translated into the other programming languages.

MATLAB is available as a program of the same name from The Mathworks, Natick, MA. The company operates a web site, `http://www.mathworks.com`, from which purchasing information is available. Many institutions have MATLAB installed on computers in computer laboratories and often make MATLAB licenses available for their members' use on personally owned computers. At the level of detail described and used here, a computer program called "GNU Octave", conceived and written by John W. Eaton (and many others), is freely available on the internet at (`http://www.gnu.org/software/octave/index.html`). It can be used to run MATLAB programs without modification.

B.2 MATLAB Files

For our purposes, the best way to use MATLAB is to use its scripting facility. With sequences of MATLAB commands contained in files, it is easy to see what calculations were done to produce a certain result, and it is easy to show that the correct values were used in producing a result. It is terribly embarrassing to produce a very nice plot that you show to your teacher or advisor only to discover later that you cannot reproduce it or anything like it for similar conditions or parameters. When the commands are in clear text files, with easily read, well-commented code, you have a very good idea of how a particular result was obtained. And you will be able to reproduce it and similar calculations as often as you please.

The MATLAB comment character is a percent sign (%). That is, lines starting with % are not read as MATLAB commands and can contain any text. Similarly, any text on a line following a % can contain textual comments and not MATLAB commands.

A MATLAB script file is a text file with the extension .m. MATLAB script files should start off with comments that identify the author, the date, and a brief description of the intent of the calculation that the file performs. MATLAB script files are invoked by typing their names without the .m at the MATLAB command line or by using their names inside another MATLAB file. Invoking the script causes the commands in the script to be executed, in order.

---

[1] For example, if the variable A represents a matrix A, its components A_{ij} are represented by A(i,j) in MATLAB and Fortran but by A[i][j] in C, C++ and Java.

MATLAB function files are also text files with the extension .m, but the first non-comment line *must* start with the word `function` and be of the form

`function` *output variable* = *function name* (*parameters*)

This defining line is called the "signature" of the function. More than one input parameter requires they be separated by commas. If a function has no input parameters, they, and the parentheses, can be omitted. Similarly, a function need not have output variables. A function can have several output variables, in which case they are separated by commas and enclosed in brackets as

`function [out1,out2,out3]=`*function name*`(in1,in2,in3,in4)`

The name of the function must be the same as the file name. Comment lines can appear either before or after the signature line, but not both, and should include the following.

(1) The first line following the signature (or the first line of the file) should repeat the signature (I often leave out the word "function") to provide a reminder of the usage of the function.
(2) Brief description of the mathematical task the function performs.
(3) Description of all the input parameters.
(4) Description of all the output parameters.

Part of the first of these lines is displayed in the "Current directory" windowpane, and the lines themselves comprise the response to the MATLAB command `help` *function name*.

The key difference between function and script files is that

• Functions are intended to be used repetitively,
• Functions can accept parameters, and,
• Variables used inside a function are invisible outside the function.

This latter point is important: variables used inside a function (except for output variables) are invisible after the function completes its tasks while variables in script files remain in the workspace.

The easiest way to produce script or function files is to use the editor packaged with the MATLAB program. Alternatively, any text editor (*e.g.*, `emacs`, `notepad`) can be used. A word processor such as Microsoft Word or

Wordpad is *not* appropriate because it embeds special formatting characters in the file and MATLAB cannot interpret them.

Because function files are intended to be used multiple times, it is a bad idea to have them print or plot things. Imagine what happens if you have a function that prints just one line of information that you think might be useful, and you put it into a loop that is executed a thousand times. Do you plan to read those lines?

MATLAB commands are sometimes terminated with a semicolon (;) and sometimes not. The difference is that the result of a calculation is printed to the screen when there is no semicolon but no printing is done when there is a semicolon. It is a good idea to put semicolons at the ends of all calculational lines in a function file. When using pseudocode presented in this book to generate MATLAB functions or scripts, you should remember to insert semicolons in order to minimize extraneous printing.

B.3 Variables, Values and Arithmetic

Values in MATLAB are usually[2] double precision numbers. When MATLAB prints values, however, it will round a number to about four digits to the right of the decimal point, or less if appropriate. Values that are integers are usually printed without a decimal point. Remember, however, that when MATLAB prints a number, it may *not* be telling you all it knows about that number.

When MATLAB prints values, it often uses a notation similar to scientific notation, but written without the exponent. For example, Avogadro's number is $6.022 \cdot 10^{23}$ in usual scientific notation, but MATLAB would display this as `6.022e+23`. The `e` denotes 10. Similarly, MATLAB would display the fraction `1/2048=4.8828e-04`. You can change the number of digits displayed with the `format` command. (See `help format` for details.)

MATLAB uses variable names to represent data. A variable name represents a matrix containing complex double-precision data. Of course, if you simply tell MATLAB `x=1`, MATLAB will understand that you mean a 1×1 matrix and it is smart enough to print x out without its decimal and imaginary parts, but make no mistake: they are there. And x can just as easily turn into a matrix.

A variable can represent some important value in a program, or it can represent some sort of dummy or temporary value. Important quantities

---

[2]It is possible to have single precision numbers or integers or other formats, but requires special declarations.

should be given names longer than a few letters, and the names should indicate the meaning of the quantity. For example, if you were using MATLAB to generate a matrix containing a table of squares of numbers, you might name the table, for example, `tableOfSquares` or `table_of_squares`.

Once you have used a variable name, it is bad practice to re-use it to mean something else. It is sometimes necessary to do so, however, and the statement

```
clear varOne varTwo
```

should be used to clear the two variables `varOne` and `varTwo` before they are re-used. This same command is critical if you re-use a variable name but intend it to have smaller dimensions.

MATLAB has a few reserved names. You should not use these as variable names in your files. If you do use such variables as `i` or `pi`, they will lose their special meaning until you clear them. Reserved names include:

ans: The result of the previous calculation.

computer: The type of computer you are on.

eps: The smallest positive number ϵ that can be represented on the computer and that satisfies the expression $1 + \epsilon > 1$. Be warned that this usage is different from the use of **eps** in the text.

i, j: The imaginary unit ($\sqrt{-1}$). Using `i` or `j` as subscripts or loop indices when you are also using complex numbers can generate incorrect answers.

inf: Infinity (∞). This will be the result of dividing 1 by 0.

NaN: "Not a Number." This will be the result of dividing 0 by 0, or **inf** by **inf**, multiplying 0 by **inf**, *etc.*

pi: π

realmax, realmin: The largest and smallest real numbers that can be represented on this computer.

version: The version of MATLAB you are running. (The **ver** command gives more detailed information.)

Arithmetic operations can be performed on variables. These operations include the following. In each case, the printed value would be suppressed if a semicolon were used.

Some MATLAB operations		
=	Assignment	`x=4` causes variable `x` to have value 4.
+	Addition	`x+1` prints the value 5.
−	Subtraction	`x-1` prints the value 3.
*	Multiplication	`2*x` prints the value 8.
/	Division	`6/x` prints the value 1.5.
^	Exponentiation	`x^3` prints the value 64.
()	Grouping	`(x+2)/2` prints the value 3.

MATLAB has a vast number of mathematical functions. MATLAB functions are called using parentheses, as in `log(5)`.

Exercise B.1. *Start up* MATLAB *or Octave and use it to answer the following questions.*

(a) *What are the values of the reserved variables* `pi`, `eps`, `realmax`, *and* `realmin`?

(b) *Use the "*`format long`*" command to display* `pi` *in full precision and "*`format short`*" to return* MATLAB *to its default, short, display.*

(c) *Set the variable* `a=1`, *the variable* `b=1+eps`, *the variable* `c=2`, *and the variable* `d=2+eps`. *What is the difference in the way that* MATLAB *displays these values?*

(d) *Do you think the values of* `a` *and* `b` *are different? Is the way that* MATLAB *formats these values consistent with your idea of whether they are different or not?*

(e) *Do you think the values of* `c` *and* `d` *are different? Explain your answer.*

(f) *Choose a value and set the variable* `x` *to that value.*

(g) *What is the square of* `x`*? Its cube?*

(h) *Choose an angle* θ *and set the variable* `theta` *to its value (a number).*

(i) *What is* $\sin\theta$*?* $\cos\theta$*? Angles can be measured in degrees or radians. Which of these has* MATLAB *used?*

B.4   Variables Are Matrices

MATLAB treats all its variables as though they were matrices. Important subclasses of matrices include row vectors (matrices with a single row and possibly several columns) and column vectors (matrices with a single column and possibly several rows). One important thing to remember is that you don't have to declare the size of your variable; MATLAB decides how big the variable is when you try to put a value in it. The easiest way to define a row vector is to list its values inside of square brackets, and separated by spaces or commas:

```
rowVector = [ 0, 1, 3, 6, 10 ]
```

The easiest way to define a column vector is to list its values inside of square brackets, separated by semicolons or line breaks.

```
columnVector1 = [ 0; 1; 3; 6; 10 ]
columnVector2 = [ 0
                  1
                  9
                  36
                  100 ]
```

(It is not necessary to line the entries up, but it makes it look nicer.) Note that `rowVector` is *not* equal to `columnVector1` even though each of their components is the same.

MATLAB has a special notation for generating a set of equally spaced values, which can be useful for plotting and other tasks. The format is:

```
start : increment : finish
```

or

```
start : finish
```

in which case the increment is understood to be 1. Both of these expressions result in row vectors. So we could define the even values from 10 to 20 by:

```
evens = 10 : 2 : 20
```

Sometimes, you'd prefer to specify the *number* of items in the list, rather than their spacing. In that case, you can use the `linspace` function, which has the form

```
linspace( firstValue, lastValue, numberOfValues )
```

in which case we could generate six even numbers with the command:

```
evens = linspace ( 10, 20, 6 )
```

or fifty evenly-spaced points in the interval [10,20] with

```
points = linspace ( 10, 20, 50 )
```

As a general rule, use the colon notation when the increment is an integer or when you know what the increment is and use `linspace` when you know the number of values but not the increment.

Another nice thing about MATLAB vector variables is that they are *flexible*. If you decide you want to add another entry to a vector, it's very easy to do so. To add the value 22 to the end of our **evens** vector:

```
evens = [ evens, 22 ]
```

and you could just as easily have inserted a value 8 before the other entries, as well.

Even though the number of elements in a vector can change, MATLAB always knows how many there are. You can request this value at any time by using the `numel` function. For instance,

```
numel ( evens )
```

should yield the value 7 (the 6 original values of 10, 12, ... 20, plus the value 22 tacked on later). In the case of matrices with more than one nontrivial dimension, the `numel` function returns the product of the dimensions. The `numel` of the empty vector is zero. The `size` function returns a vector containing two values: the number of rows and the number of columns (or the numbers along each of the dimensions for arrays with more than two dimensions). To get the number of rows of a variable v, use `size(v,1)` and to get the number of columns use `size(v,2)`. For example, since **evens** is a row vector, `size( evens, 1)=1` and `size( evens, 2)=7`, one row and seven columns.

To specify an individual entry of a vector, you need to use index notation, which uses *round* parentheses enclosing the index of an entry. *The first element of an array has index 1* (as in Fortran, but not C and Java). Thus, if you want to alter the third element of **evens**, you could say

```
evens(3) = 7
```

Exercise B.2. *Start up* MATLAB *or Octave and use it to do the following tasks:*

(a) Use the `linspace` function to create a row vector called **meshPoints** containing exactly 500 values with values evenly spaced between -1 and 1. Do not print all 500 values!

(b) What expression will yield the value of the 55th element of `meshPoints`?

(c) Use the `numel` function to confirm the vector has length 500.

(d) Produce a plot of a sinusoid on the interval $[-1, 1]$ using the command

```
plot(meshPoints,sin(2*pi*meshPoints))
```

In its very simplest form, the signature of the `plot` function is

`plot(`*array of x values, array of y values*`)`

The arrays, of course, need to have the same numbers of elements. The `plot` function has more complex forms that give you considerable control over the plot. Use `doc plot` for further documentation.

B.5 Matrix and Vector Operations

MATLAB provides a large assembly of tools for matrix and vector manipulation. The following exercise illuminates the use of these operations by example.

Exercise B.3. *Open up* MATLAB *or Octave and use it to perform the following tasks.*

Define the following vectors and matrices:

```
rowVec1 = [ -1 -4 -9]
colVec1 = [ 2
            9
            8 ]
mat1 = [ 1  3  5
         7 -9  2
         4  6  8 ]
```

(a) You can multiply vectors by constants. Compute

```
colVec2 = (pi/4) * colVec1
```

(b) The cosine function can be applied to a vector to yield a vector of cosines. Compute

```
colVec2 = cos( colVec2 )
```

Note that the values of `colVec2` have been overwritten.

(c) You can add vectors and multiply by scalars. Compute

```
colVec3 = colVec1 + 2 * colVec2
```

(d) The sum of a row vector and a column vector makes no sense in the context of this book. Old versions of MATLAB flag attempts to add row vectors to column vectors as errors, but current versions allow it. The result is a *matrix*. In the context of this book, this matrix result is *always wrong* and is a recurring source of programming errors. Compute

```
wrongMatrix = colVec1 + rowVec1;
```

Look carefully at the result. When you are testing a program, be alert for unexpected appearances of matrices and search for sums of row and column vectors that might cause these appearances!

(e) You can do row-column matrix multiplication. Compute

```
colvec4 = mat1 * colVec1
```

(f) A single quote following a matrix or vector indicates a (Hermitian) transpose.

```
mat1Transpose = mat1'
rowVec2 = colVec3'
```

Warning: The single quote means the Hermitian adjoint or complex-conjugate transpose. If you want a true transpose applied to a complex matrix you must use ".'".

(g) Transposes allow the usual operations. You might find $\mathbf{u}^T\mathbf{v}$ a useful expression to compute the dot (inner) product $\mathbf{u} \cdot \mathbf{v}$ (although there is a `dot` function in MATLAB).

```
mat2 = mat1 * mat1'      % mat2 is symmetric
rowVec3 = rowVec1 * mat1
dotProduct = colVec3' * colVec1
euclideanNorm = sqrt(colVec2' * colVec2)
```

(h) Matrix operations such as determinant and trace are available, too.

```
determinant = det( mat1 )
traceOfMat1 = trace( mat1 )
```

(i) You can pick certain elements out of a vector, too. Use the following command to find the smallest element in a vector `rowVec1`.

```
min(rowVec1)
```

(j) The **min** and **max** functions work along one dimension at a time. They produce vectors when applied to matrices.

```
max(mat1)
```

(k) You can compose vector and matrix functions. For example, use the following expression to compute the max norm of a vector.

```
max(abs(rowVec1))
```

(l) How would you find the single largest element of a matrix?
(m) As you know, a magic square is a matrix all of whose row sums, column sums and the sums of the two diagonals are the same. (One diagonal of a matrix goes from the top left to the bottom right, the other diagonal goes from top right to bottom left.) Show by direct computation that if the matrix A is given by

```
A=magic(100);   % do not print all 10,000 entries.
```

Then it has 100 row sums (one for each row), 100 column sums (one for each column) and two diagonal sums. These 202 sums should all be exactly the same, and you *could* verify that they are the same by printing them and "seeing" that they are the same. It is easy to miss small differences among so many numbers, though. *Instead*, verify that A is a magic square by constructing the 100 column sums (without printing them) and computing the maximum and minimum values of the column sums. Do the same for the 100 row sums, and compute the two diagonal sums. Check that these six values are the same. If the maximum and minimum values are the same, the flyswatter principle says that all values are the same.
Hints:

- Use the MATLAB **min** and **max** functions.
- Recall that **sum** applied to a matrix yields a row vector whose values are the sums of the columns.
- The MATLAB function **diag** extracts the diagonal of a matrix, and the composition of functions
  sum(diag(fliplr(A))) computes the sum of the other diagonal.

(n) Suppose we want a table of integers from 0 to 9, their squares and cubes. We could start with

```
integers = 0 : 9
```

but now we'll get an error when we try to multiply the entries of
`integers` by themselves.

`squareIntegers = integers * integers`

Realize that MATLAB deals with vectors, and the default multiplication
operation with vectors is row-by-column multiplication. What we want
here is *element-by-element* multiplication, so we need to place a *period*
in front of the operator:

`squareIntegers = integers .* integers`

Now we can define `cubeIntegers` and `fourthIntegers` in a similar
way.

`cubeIntegers = squareIntegers .* integers`
`fourthIntegers = squareIntegers .* squareIntegers`

Finally, we would like to print them out as a table. `integers`,
`squareIntegers`, *etc.* are row vectors, so make a matrix whose *columns*
consist of these vectors and allow MATLAB to print out the whole matrix at once.

`tableOfPowers=[integers', squareIntegers', ...`
`            cubeIntegers', fourthIntegers']`

(The "..." tells MATLAB that the command continues on the next
line.)

(o) Compute the squares of the values in `integers` alternatively using the
exponentiation operator as:

`sqIntegers = integers .^ 2`

and check that the two calculations agree with the command

`norm(sqIntegers-squareIntegers)`

that should result in zero.

(p) You can add constants to vectors and matrices. Compute

`squaresPlus1=squareIntegers+1;`

(q) Watch out when you use vectors. The multiplication, division and
exponentiation operators all have two possible forms, depending on
whether you want to operate on the arrays, or on the elements in the
arrays. In all these cases, you need to use the **period** notation to force

elementwise operations. Fortunately, as you have seen above, using multiplication or exponentiation without the dot will often produce an error. The same cannot be said of division. Compute

```
squareIntegers./squaresPlus1
```

and also

```
squareIntegers/squaresPlus1
```

This latter value uses the Moore–Penrose pseudo-inverse and is almost never what you intend. You have been warned! **Remark:** Addition, subtraction, and division or multiplication by a scalar *never* require the dot in front of the operator, although you will get the correct result if you use one.

(r) The index notation can also be used to refer to a subset of elements of the array. With the *start:increment:finish* notation, we can refer to a range of indices. Two-dimensional vectors and matrices can be constructed by leaving out some elements of our three-dimensional ones. For example, submatrices an be constructed from `tableOfPowers`. (The **end** function in MATLAB means the last value of that dimension.)

```
tableOfCubes = tableOfPowers(:,[1,3])
tableOfOddCubes = tableOfPowers(2:2:end,[1,3])
tableOfEvenFourths = tableOfPowers(1:2:end,1:3:4)
```

(s) You have already seen the MATLAB function **magic(n)**. Use it to construct a 10×10 matrix.

```
A = magic(10)
```

What commands would be needed to generate the four 5×5 matrices in the upper left quarter, the upper right quarter, the lower left quarter, and the lower right quarter of A?

Repeated Warning: Although multiplication of vectors is illegal without the dot, division of vectors is legal! It will be interpreted in terms of the Moore–Penrose pseudo-inverse. Beware!

B.6 Flow Control

It is critical to be able to ask questions and to perform repetitive calculations in m-files. These topics are examples of "flow control" constructs in programming languages. MATLAB provides two basic looping (repetition) constructs: `for` and `while`, and the `if` construct for asking questions. These statements each surround several MATLAB statements with `for`, `while` or `if` at the top and `end` at the bottom.

Remark B.1. It is an excellent idea to indent the statements between the `for`, `while`, or `if` lines and the `end` line. This indentation strategy makes code immensely more readable. Code that is hard to read is hard to debug, and debugging is hard enough as it is.

The syntax of a `for` loop is

> `for`   *control-variable=start* : *increment* : *end*
> > MATLAB *statement* ...
> > . . .
>
> `end`

The syntax of a `while` loop is

> MATLAB *statement initializing a control variable*
> `while`   *logical condition involving the control variable*
> > MATLAB *statement* ...
> > . . .
> > MATLAB *statement changing the control variable*
>
> `end`

The syntax of a simple `if` statement is

> `if`   *logical condition*
> > MATLAB *statement* ...
> > . . .
>
> `end`

The syntax of a compound `if` statement is

```
if   logical condition
   MATLAB statement ...
      ...
elseif   logical condition
      ...
else
      ...
end
```

Note that `elseif` is one word! Using two words `else if` changes the statement into two nested `if` statements with possibly a *very* different meaning, and a different number of `end` statements.

Exercise B.4. *The "max" or "sup" or "infinity" norm of a vector is given as the maximum of the absolute values of the components of the vector. Suppose $\{v_n\}_{n=1,...,N}$ is a vector in $\mathbb{R}^N$, then the infinity norm is given as*

$$\|v\|_\infty = \max_{n=1,...,N} |v_n|. \tag{B.1}$$

If v *is a* MATLAB *vector, then the* MATLAB *function* `numel` *gives its number of elements, and the following code will compute the infinity norm. Note how indentation helps make the code understandable. (*MATLAB *already has a* `norm` *function to compute norms, but this is how it could be done.)*

```
% find the infinity norm of a vector v

N = numel(v);
nrm = abs(v(1));
for n=2:N
  if abs(v(n)) > nrm
    nrm=abs(v(n));  % largest value up to now
  end
end
nrm    % no semicolon: value is printed
```

(a) Define a vector as

   `v=[ -5 2 0 6 8 -1 -7 -10 -10];`

(b) How many elements does v have? Does that agree with the result of the `numel` function?

(c) Copy the above code into the MATLAB command window and execute it.

(d) What is the first value that **nrm** takes on? (5)

(e) How many times is the statement with the comment "`largest value up to now`" executed? (3)

(f) What are all the values taken by the variable **nrm**? (5,6,7,10)

(g) What is the final value of **nrm**? (10)

B.7   Script and Function Files

If you have to type everything at the command line, you will not get very far. You need some sort of scripting capability to save the trouble of typing, to make editing easier, and to provide a record of what you have done. You also need the capability of making functions or your scripts will become too long to understand. In the next exercise, you will write a script file.

Exercise B.5.

(a) *Copy the code given above for the infinity norm into a file named* `infnrm.m`. *Recall you can get an editor window from the File→New→M-file menu or from the* `edit` *command in the command windowpane. Don't forget to save the file.*

(b) *Redefine the vector*

    v = [-35 -20 38 49 4 -42 -9 0 -44 -34];

(c) *Execute the script m-file you just created by typing just its name* (`infnrm`) *without the* `.m` *extension in the command windowpane. What is the infinity norm of this vector? (49)*

(d) *The usual Euclidean or 2-norm is defined as*

$$\|v\|_2 = \sqrt{\sum_1^N v_n^2}. \tag{B.2}$$

Copy the following MATLAB *code to compute the 2-norm into a file named* `twonrm.m`.

```
% find the two norm of a vector v
% your name and the date

N = numel(v);
nrm = v(1)^2;
for n=2:N
  nrm = nrm + v(n)^2;
```

```
end
nrm=sqrt(nrm)     % no semicolon: value is printed
```

(e) *Using the same vector* v, *execute the script* twonrm. *What are the first four values the variable* norm *takes on? (1625, 3069, 5470, 5486) What is its final value? (102.0931)*

(f) *Look carefully at the mathematical expression (B.2) and the* MATLAB *code in* twonrm.m. *The the way one translates a mathematical summation into* MATLAB *code is to follow the steps:*

   (i) *Set the initial value of the sum variable (*nrm *in this case) to zero or to the first term.*

   (ii) *Put an expression adding subsequent terms, one at a time, inside a loop. In this case it is of the form* nrm=nrm+*something.*

Script files are very convenient, but they have drawbacks. For example, if you had two different vectors, v and w, for which you wanted norms, it would be inconvenient to use infnrm or twonrm. It would be especially inconvenient if you wanted to get, for example, $\|v\|_2 + 1/\|w\|_\infty$. This inconvenience is avoided by using function m-files. Function m-files define your own functions that can be used just like MATLAB functions such as sin(x), *etc.* In the following exercise, you will write two function m-files.

Exercise B.6.

(a) *Copy the file* infnrm.m *to a file named* infnorm.m. *(Look carefully, the names are different! You can use "save as" or cut-and-paste to do the copy.) Add the following lines to the beginning of the file:*

```
function nrm = infnorm(v)
% nrm = infnorm(v)
% v is a vector
% nrm is its infinity norm
```

(b) *The first line of a function m-file is called the "signature" of the function. The first comment line repeats the signature in order to explain the "usage" of the function. Subsequent comments explain the parameters (such as v) and the output (such as* norm*) and, if possible, briefly explain the methods used. The function name and the file name must agree.*

(c) *Place a semicolon on the last line of the file so that nothing will normally be printed by the function.*

(d) Use the MATLAB *"help" command:*

```
help infnorm
```

This command will repeat the first lines of comments (up to a blank line or a line of code) and provides a quick way to refresh your memory of how the function is to be called and what it does.

(e) Invoke the function in the command windowpane by typing

```
infnorm(v)
```

(f) Repeat the above steps to define a function named twonorm.m *from the code in* twonrm.m. *Be sure to put comments in.*

(g) Define two vectors

```
a = [ -43 -37  24  27 37 ];
b = [  -5  -4 -29 -29 30 ];
```

and find the value of infinity norm of a *and the two norm of* b *with the commands*

```
aInfinity = infnorm(a)
bTwo      = twonorm(b)
```

Note that you no longer need to use the letter v *to denote the vector, and it is easy to manipulate the values of the norms.*

B.8   MATLAB Linear Algebra Functionality

MATLAB was originally conceived as a "matrix laboratory" and has considerable linear algebra functionality available. This section presents a small sample of those MATLAB functions that implement algorithms similar to those discussed in this book. Detailed instructions on use and implementation can be found in the MATLAB "help" facility that is part of the distribution package or on the Mathworks web site.

B.8.1   *Solving matrix systems in MATLAB*

MATLAB provides a collection of direct solvers for matrix systems rolled into a single command: "\". If a MATLAB variable A is an N×N matrix and b is a N×1 vector, then the solution of the system $Ax = b$ is computed in MATLAB with the command x=A\b. Although this looks unusual to a person used to mathematical notation, it is equivalent to $x = A^{-1}b$ and it respects the order of matrix operations. There is a named function, `mldivide` ("matrix left divide") that is equivalent to the symbolic operation: `x=A\b` is the identical to `x=mldivide(A,b)`. In informal speech, this capability is often called simply "backslash".

Warning:
The command `mldivide` is more general than ordinary matrix inversion. If the matrix `A` is not N×N or if `b` is not an N-vector, `mldivide` provides a *least-squares* best approximate solution, and no warning message is given. This is equivalent to the Moore–Penrose pseudoinverse. Care must be taken that typing errors do not lead to incorrect numerical results.

The underlying numerical methods for the `mldivide` command currently come from `umfpack` [Davis (2004)]. They work for both dense and sparse matrices and are among the most efficient methods known.

In addition to `mldivide` (backslash), MATLAB provides implementations of several iterative solution algorithms, only two of which are mentioned here.

`pcg` uses the CG and preconditioned CG methods.
`gmres` uses the generalized minimum residual method.

B.8.2 *Condition number of a matrix*

MATLAB has three functions to find the condition number of a matrix, using three different methods.

(1) The function `cond` computes the condition number of a matrix as presented in Definition 4.7.
(2) The function `condest` is an estimate of the 1-norm condition number.
(3) The function `rcond` is an estimate of the *reciprocal* of the condition number.

B.8.3 *Matrix factorizations*

MATLAB has functions to compute several standard matrix factorizations, as well as "incomplete" factorizations that are useful as preconditioners for iterative methods such as conjugate gradients.

`chol` computes the Cholesky factorization, an LL^t factorization for symmetric positive definite matrices.
`ichol` computes the incomplete Cholesky factorization.

lu computes the LU factorization as discussed in Section 3.6.
ilu computes the incomplete LU factorization.
qr computes the QR factorization as discussed in Section 9.6.

B.8.4 *Eigenvalues and singular values*

eig computes all the eigenvalues and eigenvectors of a matrix, as discussed
  in Chapter 9. It can handle the "generalized eigenvalue" problem also.
  It is primarily used for relatively small dense matrices.
eigs computes some of the largest or smallest eigenvalues and eigenvectors,
  or those nearest a shift, σ. It is most appropriate for large, sparse
  matrices.
svd computes the singular value decomposition of a matrix (Definition 2.5).
svds computes some of the largest singular values or those nearest a shift.
  It is most appropriate for large, sparse matrices.

B.9 Debugging

A programming error in a computer program is called a "bug". It is commonly believed that the use of the term "bug" in this way dates to a problem in a computer program that Rear Admiral Grace Hopper, [Patterson (2011)], was working with. It turned out that a moth had become trapped in a relay in the early computer, causing it to fail. This story is true and pictures of the deceased moth taped into her notebook can be found on the internet. The term "bug" in reference to errors in mechanical devices had been in use for many years at the time, but the story is so compelling that people still believe it was the source of the term.

Finding and eliminating bugs in computer programs is called "debugging". Debugging is one of the most difficult, time consuming and least rewarding activities you are likely to engage in. Software engineers teach that absolutely any programming habit you develop that reduces the likelihood of creating a bug is worth the trouble in saved debugging time. Among these habits are:

- Indenting your code inside loops and if statements;
- Use long descriptive variable names;
- Write shorter functions with fewer branches; and,
- Never reuse the same variable for two different quantities.

You are urged to adopt these and any other practices that you find help you avoid bugs.

One of the most powerful debugging tools a programmer has is a "source-level debugger", or just "debugger". MATLAB, like all other modern programming environments, incudes such a debugger, integrated into its window environment. This tool can be used to follow the execution of a MATLAB function or script line by line, by which you can understand how the code works, thereby helping to find errors. MATLAB provides an excellent tutorial on its debugger. Search the documentation for "Debugging Process and Features". If you are using another programming language, you should learn to use an appropriate debugger: the time spent learning it will be paid back manyfold as you use it.

It is beyond the scope of this book to provide tutorials on the various debuggers available for various languages. It is true, however, that there is a core functionality that all debuggers share. Some of those core functions are listed below, using MATLAB terminology. Other debuggers may have other terminology for similar functions.

Values All debuggers allow you to query the current value of variables in the current function. In MATLAB and in several other debuggers, this can be accomplished by placing the cursor over the variable and holding it stationary.

Step Execute *one line* of source code from the current location. If the line is a function call, complete the function call and continue *in the current function*.

Step in If the next line of source code is a function call, step into that function, so that the first line of the function is the line that is displayed. You would normally use this for functions you suspect contribute to the bug but not for MATLAB functions or functions you are confident are correct.

Breakpoints It is usually inconvenient to follow a large program from its beginning until the results of a bug become apparent. Instead, you set "breakpoints", which are places in the code that cause the program to stop and display source code along with values of variables. If you find a program stopping in some function, you can set a breakpoint near the beginning of that function and then track execution from that point on.

Conditional breakpoints MATLAB provides for breakpoints based on conditions. Numerical programs sometimes fail because the result of some calculation is unreasonable or unexpected. For example, if the forbidden values `inf` or `NaN`[3] are generated unexpectedly. Setting a breakpoint that will be activated as soon as such a value is generated, no matter what line of code is involved may expose the bug. It is also possible to set a breakpoint based on a condition such as `x` becoming equal to 1.

Continue Continue from the current line until the next breakpoint, or until it loops back to this breakpoint.

Call stack Most programs call many functions and often call the same function from different places. If, for example, your debugger shows that the program has just computed `inf` inside `log(x)` with `x=0`, you need to know *where* the call to `log(x)` occurred. The call stack is the list of function calls culminating with the current one.

Finally, one remarkably effective strategy to use with a debugger is to examine the source code, querying the current values of relevant variables. Then it is possible to predict the effect of the next line of code. Stepping the debugger one line will confirm your prediction or surprise you. If it surprises you, you probably have found a bug. If not, go on to the following line.

B.10   Execution Speed

The remarks in this section are specific to MATLAB and, to some extent, Octave. These remarks *cannot be generalized* to languages such as C, C++ and Fortran, although Fortran shares the array notation and use of the colon with MATLAB.

It is sometimes possible to substantially reduce execution times for some MATLAB code by reformulating it in a mathematically equivalent manner or by taking advantage of MATLAB's array notation. In this section, a few strategies are presented for speeding up programs similar to the pseudocode examples presented in this book.

The simplest timing tools in MATLAB are the `tic` and `toc` commands. These commands are used by calling `tic` just before the segment of code or function that is being timed, and `toc` just after the code is completed. The `toc` call results in the elapsed time since the `tic` call being printed. Care must be taken to place them inside a script or function file or on the same line as the code to be timed, or else it will be your typing speed that is

---

[3]Division by zero results in a special illegal value denoted `inf`. The result of 0/0 and most arithmetic performed on `inf` results in a different illegal value denoted `NaN` for "Not a Number".

measured. A second point to remember is that the first time a function is called it must be read from disk, a slow process. If you plan to measure the speed of a function, you should do it *twice*. You will find that the second value is much more reliable (and often much smaller) than the first.

B.10.1 *Initializing vectors and matrices in MATLAB*

MATLAB vectors are not fixed in length, but can grow dynamically. They do not shrink. The first time MATLAB encounters a vector, it allocates some amount of storage. As soon as MATLAB encounters an index larger than it has already allocated, it stops, allocates a new, longer, vector and (if necessary) copies all old information into the new vector. This operation involves calls to the operating system for the allocation and then (possibly) a copy. All this work can take a surprising amount of time. Passing through an array in reverse direction can avoid some of this work.

For example, on a 2012-era computer running Kubuntu Linux, the following command

```
tic; for i=1:2000; for j=1:2000; G(i,j)=i+j;end;end;toc
```

takes about 4.65 seconds. Executing the command *a second time*, so G has already been allocated, takes only 0.37 seconds. Similarly, executing the command

```
tic; for i=2000:-1:1; for j=2000:-1:1; G(i,j)=i+j;end;end;toc
```

(passing through the array in reverse order) takes 0.40 seconds. (The difference between 0.37 seconds and 0.40 seconds is not significant.)

In many computer languages, you are required to declare the size of an array *before* it is used. Such a declaration is not required in MATLAB, but a common strategy in MATLAB is to initialize a matrix to zero using the **zeros** command. It turns out that such a strategy carries a substantial advantage in computer time:

```
tic;G=zeros(2000,2000);
     for i=1:2000;for j=1:2000;G(i,j)=i+j;end;end;toc
```

(typed on a single line) takes only 0.08 seconds!

B.10.2 *Array notation and efficiency in MATLAB*

MATLAB allows arithmetic and function evaluation to be done on entire matrices at once instead of using loops. Addition, subtraction, and (row-column) multiplication can be represented in the usual manner. In addition, *componentwise* multiplication, division, exponentiation and function calls can also be done on matrices. These are summarized in Table B.1.

Table B.1   Selected MATLAB array operations.

$(A)_{ij} = a_{ij}$ for $1 \leq i \leq N_A$ and $1 \leq j \leq M_A$, similarly for B and C.

Operation	Interpretation	Restrictions
C=A+B	$c_{ij} = a_{ij} + b_{ij}$	$N_A = N_B = N_C$, $M_A = M_B = M_C$
C=A-B	$c_{ij} = a_{ij} - b_{ij}$	$N_A = N_B = N_C$, $M_A = M_B = M_C$
C=A*B	$c_{ij} = \sum_{k=1}^{M_A} a_{ik} b_{kj}$	$N_C = N_A$, $M_C = M_B$, $M_A = N_B$
C=A^n	C=A*A*A*$\cdots$*A	A is square
C=A.*B	$c_{ij} = a_{ij} * b_{ij}$	$N_A = N_B = N_C$, $M_A = M_B = M_C$
C=A./B	$c_{ij} = a_{ij}/b_{ij}$	$N_A = N_B = N_C$, $M_A = M_B = M_C$
C=A.^n	$c_{ij} = (a_{ij})^n$	$N_A = N_C$, $M_A = M_C$
C=n.^A	$c_{ij} = n^{a_{ij}}$	$N_A = N_C$, $M_A = M_C$
C=f(A)	$c_{ij} = f(a_{ij})$	$N_A = N_C$, $M_A = M_C$, f a function

Warning: A careful examination of Table B.1 shows that the expression exp(A) is *not* the same as e^A ($= \sum_{n=0}^{\infty} A^n/n!$).

The array operations described in Table B.1 are generally faster than the equivalent loops. When the matrices are large, the speed improvement for using array operations can be dramatic. On a 2012-era computer running Kubuntu Linux, a loop for adding two 4000×4000 matrices took 41 seconds, but the same matrix operation took less than 0.06 seconds!

It is often possible to optain a speedup simply by replacing a loop with equivalent array operations, even when the operations are not built-in MAT-LAB operations. For example, consider the following loop, for N=4000*4000.

```
g=zeros(N,1);
for i=1:N
  g(i)=sin(i);
end
```

This loop takes about 3.93 seconds on the computer mentioned above. A speedup of almost a factor of two is available with the simple trick of creating a vector, i=(1:N), consisting of the consecutive integers from 1 to N, as in the following code.

```
g=zeros(N,1);
i=1:N;          % i is a vector
g(i)=sin(i);    % componentwise application of sin
```

This code executes in 2.04 seconds. Once the loop has been eliminated, the code can be streamlined to pick up another 10%.

```
g=sin(1:N);
```

and this code executes in 1.82 seconds, for a total improvement of more than a factor of two.

Sometimes dramatic speed improvements are available through careful consideration of what the code is doing. The MPP2D matrix is available in MATLAB through the `gallery` function. This function provides a "rogues gallery" of matrices that can be used for testing algorithms. Recall that the MPP2D matrix is tridiagonal, and hence quite sparse. An LU factorization is available using MATLAB's `lu` function, and, given a right hand side vector b, the forward and backward substitutions can be done using `mldivide` (the "\" operator).

Consider the following code:

```
N=4000;
A=gallery('tridiag',N);
[L,U]=lu(A);
b=ones(N,1);
tic;x=U\L\b;toc
tic;y=U\(L\b);toc
```

On the same computer mentioned above, the computation of x takes 1.06 seconds, dramatically slower than the 0.0005 seconds needed to compute y. The reason is that U\L\b means the same as (U\L)\b. In this case, both U and L are bidiagonal, but (U\L) has nonzeros everywhere above the diagonal and also on the lower subdiagonal. It is quite large and it takes a long time to compute. Once it is computed, multiplying by b reduces it back to a vector. In contrast, U\(L\b) first computes the vector L\b by a simple bidiagonal multiplication and then computes the vector y with another bidiagonal multiplication.

Bibliography

Barrett, R., Berry, M., Chan, T. F., Demmel, J., Donato, J., Dongarra, J., Eijkhout, V., Pozo, R., Romine, C., and der Vorst, H. V. (1994). *Templates for the Solution of Linear Systems: Building Blocks for Iterative Methods* (SIAM, Philadelphia, PA).

Davis, T. (2004). Algorithm 832: Umfpack v4.3 — an unsymmetric-pattern multifrontal method, *ACM Transactions on Mathematical Software (TOMS)* **30**, 2, pp. 196–199.

Faber, V. and Manteuffel, T. (1984). Necessary and sufficient conditions for the existence of a conjugate gradient method, *SIAM Journal on Numerical Analysis* **21**, 2, pp. 352–362.

Gilbert, J. R., Moler, C., and Schreiber, R. (1992). Sparse matrices in matlab: Design and implementation, *SIAM Journal on Matrix Analysis and Applications* **13**, 1, pp. 333–356.

Hageman, L. A. and Young, D. M. (1981). *Applied iterative methods* (Academic Press, New York), ISBN 0123133408; 9780123133403.

Herstein, I. N. (1964). *Topics in algebra*, 1st edn. (Blaisdell Pub. Co, New York, N.Y).

Patterson, M. R. (2011). Grace murray hopper, rear admiral, united states navy, http://www.arlingtoncemetery.net/ghopper.htm.

Voevodin, V. V. (1983). The question of non-self-adjoint extension of the conjugate gradients method is closed, *USSR Computational Mathematics and Mathematical Physics* **23**, 2, pp. 143–144.

von Neumann, J. and Goldstine, H. H. (1947). Numerical inverting of matrices of high order, *Bulletin of the American Mathematical Society* **53**, 11, pp. 1021–1100.

Watkins, D. S. (1982). Understanding the qr algorithm, *SIAM Review* **24**, 4, pp. 427–440.

Index